THE COUNTRY OF T[...]
AND OTHER FICTION

SARAH ORNE JEWETT (1849–1909) grew up in South Berwick, a mill town and port on the tidal Piscataqua River in southern Maine. Her mother was Caroline Perry Jewett, her father a widely respected physician, Theodore H. Jewett, and both came from prominent and comparatively wealthy families. She completed her education under her father's tutelage and at the highly regarded Berwick Academy. Jewett's extended family included several who remembered South Berwick's most prosperous years, and they told her stories from local and family history reaching back to the American Revolution. She began writing poetry in her youth, then took up short fiction and prose sketches, publishing her first story when she was 18. In the first years of her career, she published mainly stories for girls, and she then moved gradually into sketches and stories for adult readers.

In her first book, *Deephaven* (1877), Jewett collected a group of related sketches into a unified work that received much praise. After this she settled into the pattern of her career: publishing mainly short stories in major magazines such as the *Atlantic Monthly* and at regular intervals collecting most of them into books for republication. Her most famous short story, 'A White Heron' (1886), tells of a young girl who is tempted to sacrifice a rare bird that she loves for human friendship and money. By 1890, Jewett was popular enough to be able to publish a selection of her best stories for a third time in *Tales of New England*. Though she wrote only a few novels, two have become classics. *A Country Doctor* (1884) describes a young woman's struggle to become a physician in a culture that believes women unsuitable for the profession. *The Country of the Pointed Firs* (1896) was declared an American classic by Rudyard Kipling, and many have confirmed his judgement. Her admirers have included William and Henry James, Rudyard Kipling, Oliver Wendell Holmes, John Greenleaf Whittier, and Willa Cather.

TERRY HELLER is Howard Hall Professor of English at Coe College in Cedar Rapids, Iowa. He has written *The Delights of Terror* (1987) and *The Turn of the Screw: Bewildered Vision* (1989).

THE WORLD'S CLASSICS

━━

SARAH ORNE JEWETT

The Country of the Pointed Firs

and Other Fiction

━━

Edited with an Introduction by
TERRY HELLER

Oxford   New York

OXFORD UNIVERSITY PRESS

1996

Oxford University Press, Walton Street, Oxford OX2 6DP

Oxford New York
Athens Auckland Bangkok Bombay
Calcutta Cape Town Dar es Salaam Delhi
Florence Hong Kong Istanbul Karachi
Kuala Lumpur Madras Madrid Melbourne
Mexico City Nairobi Paris Singapore
Taipei Tokyo Toronto

and associated companies in
Berlin Ibadan

Oxford is a trade mark of Oxford University Press

Editorial matter © Terry Heller 1996

First published as a World's Classics paperback 1996

British Library Cataloguing in Publication Data
Data available

Library of Congress Cataloging-in-Publication Data
Jewett, Sarah Orne, 1849–1909.
The country of the pointed firs and other fiction/Sarah Orne
Jewett; edited with an introduction by Terry Heller.
p.   cm.—(The world's classics)
Includes bibliographical references (p.   ).
(pbk.)
1. United States—Social life and customs—19th century—Fiction.
2. Maine—Social life and customs—Fiction. 3. Women—Maine—
Fiction. I. Heller, Terry, 1947– . II. Title. III. Series.
PS2132.C64 1996   813'.4—dc20   95–17532
ISBN 0–19–283190–9

1 3 5 7 9 10 8 6 4 2

Typeset by Best-set Typesetter Ltd., Hong Kong
Printed in Great Britain by
BPC Paperbacks Ltd.
Aylesbury, Bucks.

CONTENTS

ACKNOWLEDGEMENTS

I HOPE that the following list leaves out no one whose contributions to this volume were substantial: the Bostonian Society, the Boston Public Library, Matt Armstrong, Charles Cannon, Peachie Carey, Rosemary Carroll, Jane Cogie, Lisa Exey, Gina Hausknecht, David Hay, Gabriel Heller, Christine Hemmer, Peggy Knott, Sandra Kronlage, Julie Kuykendall, Dawn Offerman, Kelly Rostetter, Cody Sewell, Thomas Valley, and Susan Wolverton. Harlene Hansen, Linda Heller, Betty Rogers, and Ann Struthers helped to produce materials in the introduction and explanatory notes. Katherine Duncan-Jones identified the quotation from Sir Philip Sidney in *The Foreigner*. The *Colby Quarterly* kindly gave permission to reprint part of Ted Eden's 'A Jewett Pharmacopoeia'. Coe College provided research funds and a computer through the Howard Hall Endowed Chair, and its English Department was generous with moral support. The Stewart Memorial Library staff gave essential assistance in locating and obtaining materials and in researching the explanatory notes.

This book is dedicated to Coe College, a fine place to teach and learn.

INTRODUCTION

'Love isn't blind: it's only love that sees!'
Sarah Orne Jewett

SARAH ORNE JEWETT (1849–1909) was a popular writer at the beginning of the twentieth century. Her short stories had appeared regularly in widely read magazines such as *Atlantic Monthly*, *Harper's*, *Scribner's*, *Century*, and *McClure's* since the 1870s. Her novels had been less successful, but, except for *The Tory Lover* (1901), were well received. From its first appearance in book form in 1896, her novella *The Country of the Pointed Firs* was recognized as a masterpiece. Rudyard Kipling said of it, 'So many of the people of lesser sympathy have missed the lovely New England landscape, and the genuine breadth of heart and fun that underlies the New England nature. I maintain (and will maintain with outcries if necessary) that that is the reallest New England book ever given us.'[1] In 1915, Henry James declared Jewett to be 'mistress of an art of fiction all her own'.[2] In 1925, Willa Cather wrote in the preface to her selection of Jewett's fiction that *The Country of the Pointed Firs*, along with Nathaniel Hawthorne's *The Scarlet Letter* (1850) and Mark Twain's *Adventures of Huckleberry Finn* (1884), was an American book likely to withstand the test of time. A decade later, in 1936, Cather remembered Jewett being included in her childhood version of the card-game Authors, and she wrote that Jewett's stories stood above those of other New England writers in their 'inherent, individual beauty'. Cather went on to remark, 'The best of Jewett's work, read by a student fifty years from now, will give him the characteristic flavour, the spirit, the cadence, of an American writer of the first order.'[3] In his study of 1929, F. O. Mathiessen wrote, 'She takes her place next Emily

---

[1] Paula Blanchard, *Sarah Orne Jewett: Her World and her Work* (New York: Addison-Wesley, 1994), 304.

[2] Richard Cary, *Sarah Orne Jewett* (New York: Twayne, 1962), 29.

[3] *Not Under Forty* (New York: Knopf, 1936), 54, 78, 80.

Dickinson—the two principal women writers America has had.'[4]

After her death, Jewett's reputation suffered an eclipse. Along with many of the women writers of the United States in the nineteenth century, she was almost forgotten as the reputations of Nathaniel Hawthorne, Herman Melville, and Mark Twain ascended. Though literary scholars continued to show interest in Jewett, her readership declined, and most of her work went out of print. The causes of this shift are quite complex, but some of the main factors can be seen in Gwen Nagel's review of the reception of Jewett's work.[5] For example, Nagel's account shows that though scholars and critics who wrote about Jewett were enthusiastic, their numbers were small. One effect of this was that Jewett's work was unknown to most teachers and professors, who would depend upon their mentors and their reading to suggest texts for classroom use.

Probably a general interest in feminism and women writers among the generation of teachers that began their careers in the 1970s led to an increase in the number of students who were introduced to Jewett in the classroom. At that time one might find 'A White Heron' in the standard anthologies of United States literature, though often Jewett was not represented at all. A gradually increasing interest in women writers and a flowering of feminist scholarship in the late twentieth century helped bring about a revival in Jewett studies that included two biographies, several monographs, and Jewett's appearance in several new paperback collections and in the Library of America series. Feminists especially sought out women writers and brought them more and more to the attention of potential readers.

Nagel's summary of scholarship suggests some reasons why the numbers of Jewett's academic admirers remained small during most of the twentieth century. She points out that Jewett was labelled a local-colourist, a description that came to seem narrow, to suggest too much interest in the local and

---

[4] *Sarah Orne Jewett* (Boston: Houghton, Mifflin), 152.

[5] See Nagel's introduction to *Critical Essays on Sarah Orne Jewett* (Boston, G. K. Hall, 1984), 1–23.

particular and too little in the universal and profound. Furthermore, local colour came to be thought of as women's writing. As a result, several important women authors, notably Kate Chopin (1850–1904) and Mary Wilkins Freeman (1852–1930), were relegated along with Jewett to a minor tradition in comparison to their male contemporaries, who became known as realists and who made their reputations by writing in the more favoured literary form, the novel.

Chopin and Freeman shared with Jewett interests in writing about women's experience and in delineating a regional culture. These concerns arose in part from Jewett's admiration for the work of Harriet Beecher Stowe (1811–96), especially *The Pearl of Orr's Island* (1862), which she read as a girl. Her childhood reading also included Jane Austen and George Eliot, both of whom wrote about women's experience and emphasized regional cultures, and she later came to admire Louisa May Alcott (1832–88) as well. Jewett's desire to capture in fiction the people and places of her native Maine became a liability to her larger literary reputation, until her defenders began to point out that such an impulse is virtually universal in fiction. An emphasis on locality in fiction certainly dominated in Jewett's day, from the Scotland of Sir Walter Scott and the London of Charles Dickens to the Mississippi River of Mark Twain and the rural New England of Jewett's close friend William Greenleaf Whittier. This impulse continued into the next century in the writer she influenced most, Willa Cather (1876–1947). Cather's younger contemporary William Faulkner (1897–1962), arguably the United States' greatest novelist, memorialized his 'little postage stamp of soil' in Mississippi and produced in *Go Down, Moses* (1942) a master-work formally similar to *The Country of the Pointed Firs*. During the century after *The Country of the Pointed Firs*, some of the best United States writers continued the vital tradition to which Jewett contributed, following her in form, theme, and subject in such noteworthy works as Sherwood Anderson's *Winesburg, Ohio* (1919), Eudora Welty's *The Golden Apples* (1949), Toni Morrison's *Song of Solomon* (1977), Louise Erdrich's *Love Medicine* (1984), and Sandra Cisneros's *The House on Mango Street* (1989).

This list of works that seem related in form and theme to Jewett's work suggests one more factor that contributed to the decline of her readership after her death. Most of these books share with *The Country of the Pointed Firs* the characteristic of being novel-length works that look like collections of stories or sketches. This is a category of literary work with which critics and theorists have had some difficulty. Publishers have tended to call such works collections, to see them as somehow inferior to 'real novels', and their authors have had to assert that they were novels. Such works also tend to confuse readers who have learned to take pleasure in longer works with strong plots that include action on a fairly large scale, expansive elaboration of character, setting, and theme, and some approach to the extremes of human emotion that we associate with comedy and tragedy in their classic forms. The looser, more static, and less linear form of Jewett's masterpiece has continued to challenge readers, who often affirm their sense of the work's wholeness and completeness, but find it difficult to articulate how this has been accomplished.

In reviews and criticism of Jewett certain words are repeated that embody a paradox in how she has been evaluated: delicate, exquisite, quiet, sympathetic, sentimental, charming. Such descriptions are often meant as praise, but they also tend to belittle the stories by 'feminizing' them, by suggesting that they are marginal and of secondary importance in comparison with the robust, challenging, realistic, epic, profound, tragic, or comic—terms often applied to the better-known novels of the period. Feminist theory and criticism have tended to suggest new ways of seeing these qualities and, thereby, to emphasize Jewett's intention of presenting a female view of the world. Such an interpretation was obscured and devalued by the domination of publishing and literary studies by male writers and concerns in twentieth-century academia and especially by the shifts toward literary naturalism and political radicalism that came to the fore in American literature with the First World War and the Great Depression.

In the course of her career, Jewett discovered a clear set of aims and values, and she quite deliberately developed the skills and techniques necessary to achieve her ends. Born in

1849, she at first lived a life that was, on the whole, comfortable, secure, and quiet. She grew up in a prosperous extended family in southern Maine, in the small inland village of South Berwick, where her home is preserved as a museum. The family wealth derived mainly from commerce and shipping, South Berwick standing on a tidal estuary. Her family was highly literate, and she was encouraged to read widely. Her grandfathers were sea captains and shipowners who had travelled the world; they delighted her with tales of their adventures and enlarged her views by interesting her in far-off places and in local history. Ill health during her childhood limited her activities to some extent and nourished the habit of reading. Her delicate health and alert companionship tempted her physician father to take her with him on his rounds for the fresh air and exercise. He became her most important early teacher, emphasizing close observation of plants, natural processes, and the people who were his patients. From him she learned much about how to value her local region, its landscape, and its people, and also how to see them as representative of humanity in general. She began writing verse and stories in her childhood, and when she completed her education at the esteemed Berwick Academy in 1866, she turned to writing almost immediately, though she had considered studying medicine. She was not inclined to marry, and she was able to continue living comfortably with her family. Two years after leaving school she published her first story, 'Jenny Garrow's Lovers', a fairly conventional short romance; she continued to write and publish, and in 1869 she impressed the editors of the *Atlantic Monthly* with 'Mr Bruce', another romantic tale. Then began a sort of literary apprenticeship in which she sought guidance for improving her work from her publishers and a slowly growing circle of literary acquaintances, such as Horace Scudder, William Dean Howells, and the publisher James T. Fields, husband of Annie Fields. Early in her apprenticeship, she wrote mainly for younger readers, and she was strongly influenced by her religious beliefs and the social morality they enjoined. Her friend Theophilus Parsons, a moderate Swedenborgian Christian writer, was especially influential during these years.

As Jewett gained writing experience, her ideas about her purposes developed in ways that extended and complicated her initial impulses to tell the stories of her region and teach a positive morality to younger readers. When a young girl wrote to Jewett in 1899 to express admiration of her stories about Betty Leicester, Jewett replied with encouragement to continue reading:

you will always have the happiness of finding friendships in books, and it grows pleasanter and pleasanter as one grows older. And then the people in books are apt to make us understand 'real' people better, and to know why they do things, and so we learn sympathy and patience and enthusiasm for those we live with, and can try to help them in what they are doing, instead of being half suspicious and finding fault.[6]

In this statement, made near the end of her writing career, Jewett echoes George Eliot, who noted in chapter 17 of *Adam Bede* (1859) that while an author may not expect to reform human vices, she may hope to help readers learn to tolerate, pity, and even love their weak fellow mortals.

Jewett's main purpose became to encourage her readers to learn and cultivate the arts of friendship.[7] Related ideas appear often in Jewett's correspondence. In a letter of 1893 she wrote, 'You know there is a saying of Plato's that the best thing one can do for the people of a State is to make them acquainted with each other, and it was some instinctive feeling of this sort which led me to wish that the town and country people were less suspicious of one another.'[8] This wish arose in part out of her observations of tension between locals and the city people who began to use the Maine coast villages as summer resorts after the Civil War. The value of telling stories about ordinary life as a means of making people known to each other was something that she learnt from her father—who repeatedly advised her to tell things 'just as they are'—as well as from Stowe's *The Pearl of Orr's Island* and later from Gustave Flaubert (1821–80) and the short stories of Leo Tolstoy

    [6] *Sarah Orne Jewett: Letters*, ed. Richard Cary (Waterville, Me.: Colby College Press, 1956, 1967), 116.
    [7] Blanchard, *Sarah Orne Jewett: Her World and her Work*, 230.
    [8] *Letters*, ed. Cary, 83–4.

(1828–1910), which became models for her of 'writing about people of rustic life just as they were'.[9]

In her brief correspondence with Willa Cather in 1908, Jewett stated her aims in more aesthetic terms that echoed Nathaniel Hawthorne's world-view and foreshadowed William Faulkner's Nobel Prize acceptance speech: 'You must write to the human heart, the great consciousness that all humanity goes to make up. Otherwise what might be strength in a writer is only crudeness, and what might be insight is only observation; sentiment falls into sentimentality—you can write about life, but never write life itself.'[10]

Jewett's aims led naturally to her emphasizing certain techniques and themes. She prefigures James Joyce's *Dubliners* (1914) in her use of epiphany, the illumination or discovery of a profound and moving truth in the contemplation of ordinary events. When an epiphany leads to an intimate sharing with another person, the result is a moment of communion, the finest fruit of friendship in Jewett. Sarah W. Sherman argues that for Jewett, 'Love brings a mystical participation in another's essence. . . . These epiphanies become the source of Jewett's strength, the focal point of her life and fiction.'[11] In a letter of 1897 reflecting on the deaths of friends, Jewett wrote: 'There is something transfiguring in the best of friendship. One remembers the story of the transfiguration in the New Testament, and sees over and over in life what the great shining hours can do, and how one goes down from the mountain where they are, into the fret of everyday life again, but strong in remembrance.'[12] Again and again in Jewett's fiction, for example in 'The Life of Nancy', *The Country of the Pointed Firs*, 'The Queen's Twin', and 'Martha's Lady', the reader sees characters transfigured by friendship and strengthened for life in a difficult world.

'Fair Day' offers an example of a single epiphany as the central event. The widow Mary Bascom, vigorous at the age of

---

[9] Ibid. 8. See also Mathiessen, *Sarah Orne Jewett*, 66.

[10] *Letters of Sarah Orne Jewett*, ed. Annie Fields (Boston: Houghton, Mifflin, 1911), 249.

[11] *Sarah Orne Jewett: An American Persephone* (Hanover: University Press of New England, 1989), 64.

[12] *Letters*, ed. Fields, 126.

73, is putting her life in order, but this has been complicated by moving in with her son and daughter-in-law and away from the home farm with which she associates the most important part of her life. Another complication is that for most of her life, Mary has been estranged from Ruth Parlet, a childhood friend who later became her sister-in-law. She holds Ruth responsible for encouraging her unhappy marriage to Ruth's brother. Of special interest in this story is the indirect way in which Mary comes to the full acceptance that this quarrel must be ended. Virtually everyone in the neighbourhood has gone to the fair, so Mary takes this opportunity to visit her former home while the current tenants are gone. She finds it both familiar and attractive, but she feels locked out of it. She gradually repossesses her home by means of memory, as the familiar objects seen through a window bring back early and vivid memories. Surrounded by such memories, she finds herself put into 'a most peaceful state' (p. 207). Until now, her life has been busy, filled first with struggle and hard work until her son married, and then with the worry and trouble of being a dependant in his household. But her experience offers her a new perspective in which to view her life as a whole. From this perspective comes the epiphany, her recognition that there is a disturbing element in this peace: 'There was only one thought that would not let her be quite happy. She could not get her sister-in-law Ruth Parlet out of her mind. And strangely enough the old grudge did not present itself with the usual power of aggravation; it was of their early friendship and Ruth's good fellowship that memories would come' (p. 207). After Mary gradually adjusts herself to this revelation, she is rewarded with remembering where the house-key is probably hidden, and she is allowed to revisit the interior of her home, 'like some shell-fish finding its own old shell again and settling comfortably into the convolutions' (p. 208). So Mary learns that she would welcome a reconciliation with Ruth, even if that means taking all the blame for the quarrel on herself. And this conclusion helps her to return comfortably to her son's home.

This kind of elliptical development became increasingly important to Jewett, reaching its highest achievement perhaps in *The Country of the Pointed Firs* and the related Dunnet

Landing stories. She uses it with profoundly moving effect in 'The Queen's Twin', where the long walk of Mrs Todd and the narrator across the wild countryside to reach Abby Martin's house alters the narrator's perspective in a way that prepares her perfectly for that strange and beautiful visit. This interest in altering perspectives to create sympathy and understanding can be seen as well in 'A Neighbor's Landmark' and many other stories. Much of Jewett's most impressive fiction gains its power to move by taking the reader along with the discovering character through the seemingly commonplace incidents that change the meaning of a space and a time, making it seem sacred, or outside of normal time and space. Such eternal moments usually climax in an epiphany.

Because many of Jewett's best stories are characterized by an elliptical development, readers unacquainted with Jewett may see them as static and plotless, two of the characteristics that critics have often named as weaknesses in her work. In 'Fair Day', there is little suspense to drive the narrative forward. Mary is vaguely upset. She feels a little put-upon and rebellious and nostalgic, but the focus of these feelings is not clear. For the reader, Mary's trip home seems open-ended. Her adventures at her home do not have clear causal connections that stand out on first reading as leading to the epiphany, and yet as the story closes and the reader looks back, it seems true and right that Mary's perspective should have been altered and healed by this day's adventures. Reading the best Jewett often requires a kind of patience and trust usually demanded by fiction that seems more difficult on the surface. 'Fair Day', like 'The Honey Tree' from the end of her career, is very simply and clearly told, but attempting to explain how the story works soon makes the reader aware of its complexity.

One reason why Jewett was attracted to this kind of storytelling is reflected in her changing opinion of Jane Austen. In a letter of 1872, she complained about Austen: 'All the reasoning is done for you and all the thinking, as one might say. It seems to me like hearing somebody talk on and on and on, while you have no part in the conversation, and merely listen.'[13] When Jewett had perfected her own uses of indirection and irony, she again came to appreciate Austen. In a letter of 1902 to her closest friend, Annie Fields, she wrote, 'Yesterday afternoon I amused myself with Miss Austen's "Persuasion." Dear me, how like her people are to the people we knew years ago! . . . I am going to read another, "Persuasion" tasted so good!'[14] From the beginning of her writing career, Jewett thought it important to involve her readers in a conversation in her stories. Her indirect approach to the epiphany in 'Fair Day' shows one of the important means by which she accomplished this goal.

Jewett often extends epiphanies into shared experiences that become profound moments of communion; and her very best stories turn on such events. In 'The Life of Nancy', she tells of an unusual non-romantic friendship between Nancy Gale, an attractive rural woman, and Tom Aldis, a wealthy Boston aristocrat. They soon discover that they are kindred souls, nurtured ultimately by the beautiful Maine landscape and the sense of community that he comes to value during an extended visit to her village of East Rodney. But neither wishes to marry the other; their directions in life are not the same, and other barriers come between them and help to keep them apart for almost twenty years after they become friends. Nevertheless, when they do meet again and renew their friendship, they find its power undiminished. Their willingness at this later date to overcome all obstacles to friendship leads to a moment of communion. Finally, the only barrier to unfettered friendship between them is his regret for years of neglecting her. This problem is overcome after he has moved his family to East Rodney, when he brings Nancy, now crippled with rheumatism—a disease from which Jewett suffered, but more mildly—to visit his newly built house. Left alone with her a moment, he half voices his regret, and in doing so experiences something else:

He could not finish his sentence, for he was thinking of Nancy's long years, and the bond of friendship that absence and even forgetfulness had failed to break; of the curious insistence of fate which made him responsible for something in the life of Nancy and brought him

<hr />

[14] *Letters*, ed. Fields, 185.

back to her neighborhood. It was a moment of deep thought; he even forgot Nancy herself. He heard the water plashing on the shore below, and felt the cool sea wind that blew in at the door. (p. 301)

In this moment of silent communion, Tom moves beyond himself and Nancy to experience a kind of wonder at what appears to be a meaningful pattern in his life. Surprisingly, the landscape becomes the last stop in his mental journey outward, before Nancy's gesture calls him back again. This suggests that the landscape is the ultimate basis of their communion as well as their friendship. Tom comes to understand that his father's purchase of land in East Rodney, his friend Carew's sprained ankle, which detained Tom and so allowed him to befriend Nancy and the landscape, and a later chance meeting with Nancy in Boston—seemingly minor accidents—now seem like a 'curious insistence of fate'. Tom has helped Nancy in ways that then seemed small, but that made him responsible for something in her life. He has, in fact, helped her discover the meaningful work that makes her illness seem relatively unimportant and nourishes their friendship in his absence.

Nancy's final words in the story are not a statement of regret, spoken as they are with shining eyes, but rather a confirmation that she feels what he is feeling. She has 'read his mind' and understood his regret, but knows that it is inappropriate and unnecessary in the face of the miraculous triumph of their enduring friendship: 'There never has been a day when I haven't thought of you' (p. 301). Even in his absence, she has included him every day in her community, preserving a place for him in East Rodney. And, presumably because of Nancy, almost no one in East Rodney has forgotten him, not even new residents like the landlord at the inn where he stays on his return. Friendship is like a supernatural power, giving a pattern to what might seem random events and revealing what seems unknowable in the hearts of others.

New readers of Jewett who are unfamiliar with New England may wonder whether villages such as East Rodney are real places that they may visit as easily as Boston. In fact, Jewett rarely uses a real historical name for a village, though she often refers to quite specific settings in major cities and to other real geographical sites. The parts of 'The Life of Nancy' that are set in Boston refer to actual streets and buildings and to some historical personages in Boston after the Civil War, but when Jewett creates a village such as Dunnet Landing in *The Country of the Pointed Firs* or Ashford in 'Martha's Lady', the location is nearly always fictional. Her contemporary readers and later critics as well have speculated about models for her villages, but Jewett wished to avoid identifying most of her fictional towns with real places on the map.

   The Country of the Pointed Firs opens with what turns out to be an unusual chapter, 'The Return', which introduces the reader to the fictional village of Dunnet Landing and is the only part of the novella written in the third person. Here Jewett announces a main theme: 'The process of falling in love at first sight is as final as it is swift in such a case, but the growth of true friendship may be a life-long affair' (p. 7). The love she speaks of is not for a particular person, but for the village itself, a charming withdrawn spot on the Maine coast. The unnamed writer who becomes the narrator of the rest of the novella and of the other Dunnet Landing stories has returned to the village after a brief visit several years earlier. Love began then; friendship is about to begin. While love is instant, friendship requires time and daily effort to mature. Furthermore, as the story develops, it becomes clear that the narrator's intention in coming to stay in Dunnet for the summer is not to nurture friendship, but rather to find privacy and quiet, an escape from the bustle of Boston in order to complete a writing project. The unfolding of these and other parallel and contending impulses gives unity to the novella, but I will focus here on the way in which the narrator learns to foster intimacy.

   Dunnet Landing proves in several ways to be analogous to the herb garden of her hostess, Mrs Almira Todd, a widow of 67. The narrator soon learns that she can tell where in the garden Almira is by the odours that blow in through the window of her room. The people of Dunnet are like those herbs, and Almira is one of the keys to knowing them truly. Just as Almira's herb-gathering centres in her garden, but also spreads out to the boundaries of the community, so her acquaintance centres in the village but spreads out over a larger region. Almira introduces some of these people to the narrator; some approach her themselves as she becomes more closely associated with Almira. As Almira becomes more confident of the narrator's diplomatic skills, she presents more fragile and difficult people. The great tests of the narrator's growing skill in human agriculture come when Almira introduces the narrator to the most delicate of her friends and when the narrator ventures on her own to get to know Captain Tilley in Chapter XX, 'Along Shore'.

The narrator's progress can be measured by comparing her visits with Captain Littlepage and with Captain Tilley, both of which she makes without Mrs Todd's help. In speaking with Captain Littlepage, soon after her arrival, she is successful in drawing him out and making a friend of him, but her part of the conversation contains several moments of danger narrowly evaded and one false move that interrupts Littlepage's approach to the story of 'the waiting place' and requires the narrator to bring him carefully back to this confidence. With Captain Tilley, near the end of her stay, she makes no mistakes. The chapters between these two visits consist mainly of meetings and conversations with new people, in which the narrator refines her abilities to converse until Almira bestows her highest praise on her. When a prized visitor, Mrs Fosdick, complains that 'Conversation's got to have some root in the past, or else you've got to explain every remark you make,' Almira replies, referring to the narrator, 'Yes'm, old friends is always best, 'less you can catch a new one that's fit to make an old one out of' (p. 57). This blessing is extended when, at the Bowden reunion, Almira in effect makes the narrator an honorary member of her family by serving her a piece of apple pie upon which is inscribed 'the whole word *Bowden*' (p. 98).

What are the skills the narrator learns and refines during this period? There are many that can be identified and examined, but most can be grouped under the labels of tact and sympathy. The model of these abilities is Mrs Blackett,

Almira's aged and universally admired mother. In Chapter X, 'Where Pennyroyal Grew', the narrator sees into Mrs Blackett's power:

> she had the gift which so many women lack, of being able to make themselves and their houses belong entirely to a guest's pleasure,— that charming surrender for the moment of themselves and whatever belongs to them, so that they make a part of one's own life that can never be forgotten. Tact is after all a kind of mind-reading, and my hostess held the golden gift. Sympathy is of the mind as well as the heart, and Mrs Blackett's world and mine were one from the moment we met. Besides, she had that final, that highest gift of heaven, a perfect self-forgetfulness. (p. 45)

Mrs Blackett knows herself well enough to be secure in the kind of self-surrender that involves giving oneself wholly and openly to another. To do this is to make oneself and the moment transcendent and unforgettable. This surrender arises as well from confidence in the other person, which in turn arises from and leads to an imaginative union with that person. Not simply a sentimental identification with the other's feelings, this power includes rightly guessing another's thoughts; it embraces reading the mind as well as the heart.

What Jewett means by 'perfect self-forgetfulness' may be somewhat puzzling, but she gives Nancy Gale in 'The Life of Nancy' the same quality. It seems to be the ability to rise above one's personal desires and prejudices in the interest of sustaining friendship. Mrs Blackett's perfection of this ability shows most clearly perhaps at the Bowden reunion, where Almira sometimes and other people often let personal wishes and complaints endanger the general sense of family harmony and closeness that can produce communion there. At one point Almira turns Mrs Caplin away from her wish to gossip about Santin Bowden's alcoholism, but then falls into criticizing Marie Harris, a woman she particularly dislikes. At that moment, Mrs Blackett appears to see that her daughter is 'out of mischief' (p. 94) and, with the help of the narrator, turns the conversation from Marie's shortcomings to ideas for renewing neglected friendships.

Mrs Blackett more than once calls people away from petty difference and towards unity. Those who follow her example experience most fully the meaning of the reunion feast that the narrator understands when they march across a field to the grove where it is held: 'we were no more a New England family celebrating its own existence and simple progress; we carried the tokens and inheritance of all such households from which this had descended, and were only the latest of our line' (pp. 91–2). Mrs Blackett's skills of friendship invite all who will emulate her into communion with her family, a communion that to a significant degree transcends time and space. At the end of the narrator's first meeting with Mrs Blackett in Chapter XI, 'The Old Singers', such a moment is achieved when Mrs Blackett invites the narrator to sit in her quilted rocking-chair. The narrator sits for a few minutes and sees the centre of Mrs Blackett's world much as she herself sees it. This moment of silence leads to each imaginatively identifying with the other: 'I looked up, and we understood each other without speaking. "I shall like to think o' your settin' here today," said Mrs Blackett' (p. 51). Each is from this moment indelibly a part of the other, her communion and community enlarged.

The importance of epiphany and communion in Jewett's works shows that for her, the most significant human events occur on a spiritual level. She often said she believed that ultimately her writing came from the inexpressible in her own heart and soul, and perhaps from a larger spiritual source. 'The great messages and discoveries of literature come to us, they *write* us, and we do not control them,' Jewett said in a letter of 1885. [15] In a letter to Fields, she said of another story, 'For two weeks I have been noticing a certain string of things and having hints of character, etc., and day before yesterday the plan of the story comes into my mind, and in half an hour I have put all the little words and ways into their places and can read it off to myself like print. Who does it? for I grow more and more sure that I don't!'[16] Here, as in her comments on how characters in good stories help us to love better those we live among, and in her admonition to Cather to write to the human heart, Jewett seems to be speaking of a spiritual

[15] *Letters*, ed. Cary, 52.   [16] *Letters*, ed. Fields, 52.

level of human experience from which great literature comes and towards which it points. It is probably to this spiritual level that she refers in a letter written (in 1894 or 1895) to Sarah Whitman in response to her appreciation of 'Martha's Lady': '[I]t is those unwritable things that the story holds in its heart, if it has any, that make the true soul of it, and these must be understood.'[17]

By emphasizing epiphany and communion in Jewett's stories, I may easily leave the impression that she does not attend to the darker sides of human experience. While it is true that most of Jewett's stories, including those in this selection, tend to come to affirmative, sometimes comic resolutions, Jewett nevertheless makes her readers well aware that the struggles for communion, love, understanding, and individual fulfilment are rarely easy. In the metaphor of the poet Ann Struthers, Jewett writes 'bold yellow sentences on the dark'.[18] Though Jewett does not often put them at the centres of her stories, her fictional world is well filled with ambitious young people imprisoned in restrictive country lives, with puritanical moralists who wring the joy out of their neighbours' lives and their own too, and with the lost who for various reasons are cut off from love and community. Even at the noble feast of the Bowden reunion, the narrator of *The Country of the Pointed Firs* reflects sadly upon how much of human potential is lost: 'I was full of wonder at the waste of human ability in this world, as a botanist wonders at the wastefulness of nature, the thousands of seeds that die, the unused provision of every sort' (p. 97). Furthermore, Jewett's strong and positive characters often live with real adversities such as loneliness, poverty, disease, and oppression.

Perhaps in part because Jewett was not notably a participant in an age of reform that included the women's suffrage movement, her feminist concerns remained mainly unnoticed until feminist critics began to study her work. Her fiction shows her to have been intensely aware of the anomalous position of women in a patriarchal society. The best-known example is *A*

---

[17] *Letters*, ed. Fields, 112.
[18] *The Alcott Family Arrives* (Cedar Rapids, Ia.: Coe Review Press, 1993), 77.

Country Doctor (1884), a novel that explores the difficulties a young woman encounters when she decides to become a physician. Jewett also wrote a number of stories about women suffering because they lack power. In 'A Neighbor's Landmark', Lizzie Packer cannot move her father to preserve the landmark trees that he is determined to sell just to show that he has the power to do so and that he rules in his family. Jewett also wrote about women who are triumphant in self-realization despite their relative lack of power, notably Betsey Lane and Nancy Gale. Another important feminist theme in 'The Life of Nancy' is the exploration of friendship between men and women. Here she gives special attention to conventional expectations that such friendships must become sexual. Tom Aldis deprives himself and Nancy of years of companionship because he fears a complicating sexual relationship with her.

Once the reader begins to see how Jewett's stories of women's lives comment knowingly on the limitations imposed by patriarchal assumptions, he or she will perceive feminist themes in nearly all of her stories. Furthermore, as critics such as Ammons, Sherman, Mobley, and Roman have shown, the complexity and range of Jewett's feminist project go well beyond any simple exploration of themes. Jewett's mode of storytelling and the overall vision of her fiction reveal an intention to revise the ways her readers see and to challenge and subvert the ideology that marginalizes women, their concerns, and their power.

Though few of Jewett's stories are tragic, the realistic undercurrent of tragedy and loss is nearly always present. Anger at the waste of women's energy in her society runs near the surface in 'A Neighbor's Landmark', but it hardens into delicious satire in a story that Kipling especially admired, 'The Guests of Mrs Timms', where comic punishment falls upon a sanctimonious social climber.

'The Guests of Mrs Timms' helps to establish Jewett as one of the more accomplished American comic writers of her century. Though her wit, humour, and irony are usually subtle and are often subordinated to other more serious intentions, she is nevertheless, as Cather suggested, not unworthy of comparison with Mark Twain. Indeed, while other American women were writing fine humour in the nineteenth century—Marietta Holley (1826–1936), Alcott, Freeman, and Stowe among them—one would be hard pressed to find another so accomplished in humour as Jewett.

In this selection are many delightful comedies, from the broad humour of 'Bold Words at the Bridge', through the subtle and gentle whimsy of 'The Queen's Twin' and William Blackett's courtship in 'A Dunnet Shepherdess', to the triumphant comedy of Betsey's resurrection in 'The Flight of Betsey Lane'. Willa Cather saw Jewett's humour especially in her use of the Maine vernacular, in 'pithy bits of local speech' that she handled with delicacy and tact, and in 'her own fine attitude toward her subject-matter'. Describing that attitude in her preface to *The Country of the Pointed Firs*, Cather went on to comment, 'She had with her own stories and her own characters a very charming relation; spirited, gay, tactful, noble in its essence and a little arch in its expression.' Cather aptly describes not only Jewett's relation with her stories and characters, but the relation that she creates between her readers and her works. Though the satire can bite, as it does in 'The Guests of Mrs Timms', Jewett's humour is nearly always redemptive, shedding a genial, gently corrective light on human failings.

In the year before Jewett died, Cather met Jewett and corresponded with her. One result was a remarkable pair of letters in which Jewett advised Cather, then in her early thirties, to abandon her successful journalistic career and give all her energy to writing fiction. Cather took Jewett's advice, not only about changing her career, but also about turning her artistic attention away from attempts to emulate Henry James and towards the personal experience that she knew best. Within five years, she published *O Pioneers!* (1913), a novel that shows Jewett's influence in a variety of ways, from its opening dedication to Jewett to several elements in its plot and themes, notably the device of the man who leaves a remarkable woman and a landscape that he identifies with her and then returns after a long time, better able to value both. Jewett uses this pattern in 'The Life of Nancy' among other stories, and Cather was to use it at least one more time in one of her masterpieces, *My Ántonia* (1918).

When in 1925 Cather wrote her preface to *The Country of the Pointed Firs*, she remarked, 'I like to think with what pleasure, with what a sense of rich discovery, the young student of American literature in far distant years to come will take up this book and say, "A masterpiece!" as proudly as if he himself had made it.' In precisely crafted stories, Jewett offers a mature feminine perspective on the culture of her time and place, while speaking in true pictures to the human heart of how we can learn to love each other better. The result is, in Cather's words, 'inherent, individual beauty'.

NOTE ON THE TEXTS

The Country of the Pointed Firs first appeared in four parts in the *Atlantic Monthly* in 1896. Jewett revised it fairly extensively for book publication. The text reprinted here is from the first edition. Jewett usually revised her stories for book publication and carefully oversaw them as they went to press. The following list indicates the original magazine publications and the sources for the texts of the other stories in this volume.

'The Queen's Twin': *Atlantic Monthly* (February 1899); reprinted from *The Queen's Twin and Other Stories*, 1899.

'A Dunnet Shepherdess': *Atlantic Monthly* (December 1899); reprinted from *The Queen's Twin and Other Stories*, 1899.

'The Foreigner': reprinted from *Atlantic Monthly* (August 1900).

'William's Wedding': reprinted from *Atlantic Monthly* (August 1910). This story was left unfinished at Jewett's death.

'Fair Day': *Scribner's Magazine* (August 1888); reprinted from *Strangers and Wayfarers*, 1890.

'The Flight of Betsey Lane': *Scribner's Magazine* (August 1893); reprinted from *A Native of Winby and Other Tales*, 1893.

'The Only Rose': *Atlantic Monthly* (January 1894); reprinted from *The Life of Nancy*, 1895.

'The Guests of Mrs Timms': *Century Magazine* (September 1894); reprinted from *The Life of Nancy*, 1895.

'A Neighbor's Landmark': *Century Magazine* (December 1894); reprinted from *The Life of Nancy*, 1895.

'The Life of Nancy': *Atlantic Monthly* (February 1895); reprinted from *The Life of Nancy*, 1895.

'Martha's Lady': *Atlantic Monthly* (October 1897); reprinted from *The Queen's Twin and Other Stories*, 1899.

'Bold Words at the Bridge': *McClure's Magazine* (April 1899); reprinted from *The Queen's Twin and Other Stories*, 1899.

'The Honey Tree': reprinted from *Harper's Magazine* (December 1901).

SELECT BIBLIOGRAPHY

Works of Sarah Orne Jewett

Deephaven (Boston: Osgood and Co., 1877).

Old Friends and New (Boston: Osgood and Co., 1879).

Country By-Ways (Boston: Houghton, Mifflin, 1881).

The Mate of the Daylight, and Friends Ashore (Boston: Houghton, Mifflin, 1884).

A Country Doctor (Boston: Houghton, Mifflin, 1884).

A Marsh Island (Boston: Houghton, Mifflin, 1885).

A White Heron and Other Stories (Boston: Houghton, Mifflin, 1886).

The King of Folly Island and Other People (Boston: Houghton, Mifflin, 1888).

Strangers and Wayfarers (Boston: Houghton, Mifflin, 1890).

Tales of New England (Boston: Houghton, Mifflin, 1890).

A Native of Winby and Other Tales (Boston: Houghton, Mifflin, 1893).

The Life of Nancy (Boston: Houghton, Mifflin, 1895).

The Country of the Pointed Firs (Boston: Houghton, Mifflin, 1896).

The Queen's Twin and Other Stories (Boston: Houghton, Mifflin, 1899).

The Tory Lover (Boston: Houghton, Mifflin, 1901).

Letters of Sarah Orne Jewett, ed. Annie Fields (Boston: Houghton, Mifflin, 1911).

Sarah Orne Jewett: Letters, ed. Richard Cary, 2nd edn. (Waterville, Me.: Colby College Press, 1967).

The Uncollected Short Stories of Sarah Orne Jewett, ed. Richard Cary (Waterville, Me.: Colby College Press, 1971).

Jewett Novels and Stories, ed. Michael Bell (New York: Penguin, Library of America, 1994).

Works about Sarah Orne Jewett

Ammons, Elizabeth, 'Going in Circles: The Female Geography of *The Country of the Pointed Firs*', *Studies in the Literary Imagination*, 16 (autumn 1983), 83–92.

Blanchard, Paula, *Sarah Orne Jewett: Her World and her Work* (New York: Addison-Wesley, 1994).

Cary, Richard (ed.), *Appreciation of Sarah Orne Jewett: 29 Interpretive Essays* (Waterville, Me.: Colby College Press, 1973).

—— *Sarah Orne Jewett* (New York: Twayne, 1962).

Cather, Willa, *Not Under Forty* (New York: Knopf, 1936).

Donovan, Josephine, 'A Woman's Vision of Transcendence: A New Interpretation of the Works of Sarah Orne Jewett', *Massachusetts Review*, 21 (1980), 365–80.

—— *Sarah Orne Jewett* (New York: Ungar, 1962).

Howard, June (ed.), *New Essays on The Country of the Pointed Firs* (London: Cambridge University Press, 1994).

Matthiessen, F. O., *Sarah Orne Jewett* (Boston: Houghton, Mifflin, 1929).

Mobley, M. S., *Folk Roots and Mythic Wings in Sarah Orne Jewett and Toni Morrison* (Baton Rouge: Louisiana State University Press, 1991).

Morrison, Jane, with Peter Namuth and Cynthia Keyworth, *Master Smart Woman: A Portrait of Sarah Orne Jewett* (Unity, Me.: North Country Press, 1987).

Nagel, Gwen L. (ed.), *Critical Essays on Sarah Orne Jewett* (Boston: G. K. Hall, 1984).

—— '*Sarah Orne Jewett: A Reference Guide*, An Update', *American Literary Realism*, 17: 2 (1984), 228–63.

Nagel, Gwen and James, *Sarah Orne Jewett: A Reference Guide* (Boston: G. K. Hall, 1978).

Pratt, Annis, 'Women and Nature in Modern Fiction', *Contemporary Literature*, 13 (1972), 476–90.

Roman, Margaret, *Sarah Orne Jewett: Reconstructing Gender* (Tuscaloosa: University of Alabama Press, 1992).

Sherman, Sarah W., *Sarah Orne Jewett: An American Persephone* (Hanover: University Press of New England, 1989).

Silverthorne, Elizabeth, *Sarah Orne Jewett: A Writer's Life* (New York: Overlook Press, 1993).

Subbaraman, Sivagami, 'Rites of Passage: Narratorial Plurality as Structure in Jewett's *The Country of the Pointed Firs*', *Centennial Review*, 33: 1 (1989), 60–74.

Weber, Clara Carter and Carl J., *A Bibliography of the Published Writings of Sarah Orne Jewett* (Waterville, Me.: Colby College Press, 1949).

Zagarell, Sandra, 'Narrative of Community: The Identification of a Genre', *Signs*, 13: 3 (1988), 498–527.

A CHRONOLOGY OF
THEODORA SARAH ORNE JEWETT

1849 (3 September) Born in South Berwick, Maine, the second of three daughters of Dr Theodore H. Jewett, a physician, and Caroline Perry Jewett. She is educated at local schools, privately by her father, and at the Berwick Academy.

1860 Paternal grandfather, Theodore F. Jewett, dies.

1866 Graduates from Berwick Academy, completing her formal education.

1868 'Jenny Garrow's Lovers', her first story, is published under the name A. C. Eliot, which echoes the pseudonym of one of her favourite writers, Mary Ann Evans (George Eliot).

1869 Jewett's first publication in a major magazine, *Atlantic Monthly*: 'Mr Bruce'.

1877 Publication of her first book, *Deephaven*, a collection of related sketches.

1878 (20 September) Her father dies of a heart attack.

1880 Begins an important lifelong friendship with a Boston literary figure, Annie Fields (1834–1915).

1881 Death of the publisher and editor James T. Fields, husband of Annie Fields. Thereafter, Fields and Jewett live together during parts of each year and travel widely in Europe and America. At Fields's home in Boston, Jewett meets contemporary literary and intellectual figures. These meetings lead to close friendships with many writers, including Oliver Wendell Holmes, William Dean Howells, Henry James, Rudyard Kipling, and John Greenleaf Whittier.

1882 First trip to Europe with Annie Fields.

1884 *A Country Doctor*, a novel, published. Dr Leslie, the title character, is based on Jewett's father.

1891 (21 October) Her mother dies after a long illness.

1892 Second trip to Europe.

1896 Cruises among the Caribbean islands; publishes *The Country of the Pointed Firs* to wide acclaim, receiving letters of praise from William James and Rudyard Kipling.

1897 Her younger sister, Caroline Jewett Eastman, dies.

1898 Third trip to Europe.

1900 Fourth trip to Europe.

1901 Jewett becomes the first woman to receive an honorary Doctor of Letters degree from Bowdoin College.

1902 Injured in a carriage accident on her birthday, she is incapacitated for the rest of her life. Though she continues writing letters and visiting, she cannot sustain the necessary attention for writing fiction.

1904 Publication of 'A Spring Sunday', the last story published in her lifetime.

1908 Meets and corresponds with Willa Cather, giving her advice about her career that is generally seen as altering its direction. As a result, Cather gradually abandons journalism to devote herself to fiction. Cather subsequently dedicates her first Nebraska novel, *O Pioneers!* (1913), 'To the memory of Sarah Orne Jewett in whose beautiful and delicate work there is the perfection that endures'. Jewett's last work published in her lifetime is her poem 'The Gloucester Mother'.

1909 (24 June) Dies of cerebral haemorrhage at her home in South Berwick.

THE COUNTRY OF
THE POINTED FIRS

To

ALICE GREENWOOD HOWE*

CONTENTS

CONTENTS

I

THE RETURN

THERE was something about the coast town of Dunnet which
made it seem more attractive than other maritime villages of
eastern Maine. Perhaps it was the simple fact of acquaintance
with that neighborhood which made it so attaching, and gave
such interest to the rocky shore and dark woods, and the few
houses which seemed to be securely wedged and tree-nailed
in among the ledges by the Landing. These houses made the
most of their seaward view, and there was a gayety and determined floweriness in their bits of garden ground; the small-
paned high windows in the peaks of their steep gables were
like knowing eyes that watched the harbor and the far sea-line
beyond, or looked northward all along the shore and its
background of spruces and balsam firs. When one really
knows a village like this and its surroundings, it is like becoming acquainted with a single person. The process of falling in
love at first sight is as final as it is swift in such a case, but the
growth of true friendship may be a life-long affair.

After a first brief visit made two or three summers before in
the course of a yachting cruise, a lover of Dunnet Landing
returned to find the unchanged shores of the pointed firs, the
same quaintness of the village with its elaborate conventionalities; all that mixture of remoteness, and childish certainty
of being the centre of civilization of which her affectionate
dreams had told. One evening in June, a single passenger
landed upon the steamboat wharf. The tide was high, there
was a fine crowd of spectators, and the younger portion of the
company followed her with subdued excitement up the narrow street of the salt-aired, white-clapboarded little town.

II

MRS TODD

LATER, there was only one fault to find with this choice of a summer lodging-place, and that was its complete lack of seclusion. At first the tiny house of Mrs Almira Todd, which stood with its end to the street, appeared to be retired and sheltered enough from the busy world, behind its bushy bit of a green garden, in which all the blooming things, two or three gay hollyhocks and some London-pride, were pushed back against the gray-shingled wall. It was a queer little garden and puzzling to a stranger, the few flowers being put at a disadvantage by so much greenery; but the discovery was soon made that Mrs Todd was an ardent lover of herbs, both wild and tame, and the sea-breezes blew into the low end-window of the house laden with not only sweet-brier and sweet-mary, but balm and sage and borage and mint, wormwood and southernwood. If Mrs Todd had occasion to step into the far corner of her herb plot, she trod heavily upon thyme, and made its fragrant presence known with all the rest. Being a very large person, her full skirts brushed and bent almost every slender stalk that her feet missed. You could always tell when she was stepping about there, even when you were half awake in the morning, and learned to know, in the course of a few weeks' experience, in exactly which corner of the garden she might be.

At one side of this herb plot were other growths of a rustic pharmacopoeia, great treasures and rarities among the commoner herbs. There were some strange and pungent odors that roused a dim sense and remembrance of something in the forgotten past. Some of these might once have belonged to sacred and mystic rites, and have had some occult knowledge handed with them down the centuries; but now they pertained only to humble compounds brewed at intervals with molasses or vinegar or spirits in a small caldron on Mrs Todd's kitchen stove. They were dispensed to suffering neighbors, who usually came at night as if by stealth, bringing their

own ancient-looking vials to be filled. One nostrum was called
the Indian remedy, and its price was but fifteen cents; the
whispered directions could be heard as customers passed the
windows. With most remedies the purchaser was allowed to
depart unadmonished from the kitchen, Mrs Todd being a
wise saver of steps; but with certain vials she gave cautions,
standing in the doorway, and there were other doses which
had to be accompanied on their healing way as far as the gate,
while she muttered long chapters of directions, and kept up
an air of secrecy and importance to the last. It may not have
been only the common ails of humanity with which she tried
to cope; it seemed sometimes as if love and hate and jealousy
and adverse winds at sea might also find their proper remedies among the curious wild-looking plants in Mrs Todd's
garden.

The village doctor and this learned herbalist were upon the
best of terms. The good man may have counted upon the
unfavorable effect of certain potions which he should find his
opportunity in counteracting; at any rate, he now and then
stopped and exchanged greetings with Mrs Todd over the
picket fence. The conversation became at once professional
after the briefest preliminaries, and he would stand twirling a
sweet-scented sprig in his fingers, and make suggestive jokes,
perhaps about her faith in a too persistent course of thoroughwort elixir, in which my landlady professed such firm
belief as sometimes to endanger the life and usefulness of
worthy neighbors.

To arrive at this quietest of seaside villages late in June,
when the busy herb-gathering season was just beginning, was
also to arrive in the early prime of Mrs Todd's activity in the
brewing of old-fashioned spruce beer. This cooling and refreshing drink had been brought to wonderful perfection
through a long series of experiments; it had won immense
local fame, and the supplies for its manufacture were always
giving out and having to be replenished. For various reasons,
the seclusion and uninterrupted days which had been looked
forward to proved to be very rare in this otherwise delightful
corner of the world. My hostess and I had made our shrewd
business agreement on the basis of a simple cold luncheon at

noon, and liberal restitution in the matter of hot suppers, to
provide for which the lodger might sometimes be seen hurrying down the road, late in the day, with cunner line* in hand.
It was soon found that this arrangement made large allow-
ance for Mrs Todd's slow herb-gathering progresses through
woods and pastures. The spruce-beer customers were pretty
steady in hot weather, and there were many demands for
different soothing syrups and elixirs with which the unwise
curiosity of my early residence had made me acquainted.
Knowing Mrs Todd to be a widow, who had little beside this
slender business and the income from one hungry lodger to
maintain her, one's energies and even interest were quickly
bestowed, until it became a matter of course that she should
go afield every pleasant day, and that the lodger should
answer all peremptory knocks at the side door.

In taking an occasional wisdom-giving stroll in Mrs Todd's
company, and in acting as business partner during her frequent absences, I found the July days fly fast, and it was not
until I felt myself confronted with too great pride and pleasure in the display, one night, of two dollars and twenty-seven
cents which I had taken in during the day, that I remembered
a long piece of writing, sadly belated now, which I was bound
to do. To have been patted kindly on the shoulder and called
'darlin',' to have been offered a surprise of early mushrooms
for supper, to have had all the glory of making two dollars and
twenty-seven cents in a single day, and then to renounce it all
and withdraw from these pleasant successes, needed much
resolution. Literary employments are so vexed with uncertainties at best, and it was not until the voice of conscience
sounded louder in my ears than the sea on the nearest pebble
beach that I said unkind words of withdrawal to Mrs Todd.
She only became more wistfully affectionate than ever in her
expressions, and looked as disappointed as I expected when I
frankly told her that I could no longer enjoy the pleasure of
what we called 'seein' folks.' I felt that I was cruel to a whole
neighborhood in curtailing her liberty in this most important
season for harvesting the different wild herbs that were so
much counted upon to ease their winter ails.

'Well, dear,' she said sorrowfully, 'I've took great advantage o' your bein' here. I ain't had such a season for years, but I have never had nobody I could so trust. All you lack is a few qualities, but with time you'd gain judgment an' experience, an' be very able in the business. I'd stand right here an' say it to anybody.'

Mrs Todd and I were not separated or estranged by the change in our business relations; on the contrary, a deeper intimacy seemed to begin. I do not know what herb of the night it was that used sometimes to send out a penetrating odor late in the evening, after the dew had fallen, and the moon was high, and the cool air came up from the sea. Then Mrs Todd would feel that she must talk to somebody, and I was only too glad to listen. We both fell under the spell, and she either stood outside the window, or made an errand to my sitting-room, and told, it might be very commonplace news of the day, or, as happened one misty summer night, all that lay deepest in her heart. It was in this way that I came to know that she had loved one who was far above her.

'No, dear, him I speak of could never think of me,' she said. 'When we was young together his mother didn't favor the match, an' done everything she could to part us; and folks thought we both married well, but 't wa'n't what either one of us wanted most; an' now we're left alone again, an' might have had each other all the time. He was above bein' a sea-farin' man, an' prospered more than most; he come of a high family, an' my lot was plain an' hard-workin'. I ain't seen him for some years; he's forgot our youthful feelin's, I expect, but a woman's heart is different; them feelin's comes back when you think you've done with 'em, as sure as spring comes with the year. An' I've always had ways of hearin' about him.'

She stood in the centre of a braided rug, and its rings of black and gray seemed to circle about her feet in the dim light. Her height and massiveness in the low room gave her the look of a huge sibyl, while the strange fragrance of the mysterious herb blew in from the little garden.

III

THE SCHOOLHOUSE

FOR some days after this, Mrs Todd's customers came and went past my windows, and, haying-time being nearly over, strangers began to arrive from the inland country, such was her widespread reputation. Sometimes I saw a pale young creature like a white windflower left over into midsummer, upon whose face consumption had set its bright and wistful mark; but oftener two stout, hard-worked women from the farms came together, and detailed their symptoms to Mrs Todd in loud and cheerful voices, combining the satisfactions of a friendly gossip with the medical opportunity. They seemed to give much from their own store of therapeutic learning. I became aware of the school in which my landlady had strengthened her natural gift; but hers was always the governing mind, and the final command, 'Take of hy'sop one handful' (or whatever herb it was), was received in respectful silence. One afternoon, when I had listened,—it was impossible not to listen, with cottonless ears,—and then laughed and listened again, with an idle pen in my hand, during a particularly spirited and personal conversation, I reached for my hat, and, taking blotting-book and all under my arm, I resolutely fled further temptation, and walked out past the fragrant green garden and up the dusty road. The way went straight uphill, and presently I stopped and turned to look back.

The tide was in, the wide harbor was surrounded by its dark woods, and the small wooden houses stood as near as they could get to the landing. Mrs Todd's was the last house on the way inland. The gray ledges of the rocky shore were well covered with sod in most places, and the pasture bayberry and wild roses grew thick among them. I could see the higher inland country and the scattered farms. On the brink of the hill stood a little white schoolhouse, much wind-blown and weather-beaten, which was a landmark to seagoing folk; from its door there was a most beautiful view of sea and shore. The summer vacation now prevailed, and after finding the door unfastened, and taking a long look through one of the seaward windows, and reflecting afterward for some time in a shady place near by among the bayberry bushes, I returned to the chief place of business in the village, and, to the amusement of two of the selectmen, brothers and autocrats of Dunnet Landing, I hired the schoolhouse for the rest of the vacation for fifty cents a week.

Selfish as it may appear, the retired situation seemed to possess great advantages, and I spent many days there quite undisturbed, with the sea-breeze blowing through the small, high windows and swaying the heavy outside shutters to and fro. I hung my hat and luncheon-basket on an entry nail as if I were a small scholar, but I sat at the teacher's desk as if I were that great authority, with all the timid empty benches in rows before me. Now and then an idle sheep came and stood for a long time looking in at the door. At sundown I went back, feeling most businesslike, down toward the village again, and usually met the flavor, not of the herb garden, but of Mrs Todd's hot supper, halfway up the hill. On the nights when there were evening meetings or other public exercises that demanded her presence we had tea very early, and I was welcomed back as if from a long absence.

Once or twice I feigned excuses for staying at home, while Mrs Todd made distant excursions, and came home late, with both hands full and a heavily laden apron. This was in pennyroyal time, and when the rare lobelia was in its prime and the elecampane was coming on. One day she appeared at the schoolhouse itself, partly out of amused curiosity about my industries; but she explained that there was no tansy in the neighborhood with such snap to it as some that grew about the schoolhouse lot. Being scuffed down all the spring made it grow so much the better, like some folks that had it hard in their youth, and were bound to make the most of themselves before they died.

IV

AT THE SCHOOLHOUSE WINDOW

ONE day I reached the schoolhouse very late, owing to attendance upon the funeral of an acquaintance and neighbor, with whose sad decline in health I had been familiar, and whose last days both the doctor and Mrs Todd had tried in vain to ease. The services had taken place at one o'clock, and now, at quarter past two, I stood at the schoolhouse window, looking down at the procession as it went along the lower road close to the shore. It was a walking funeral, and even at that distance I could recognize most of the mourners as they went their solemn way. Mrs Begg had been very much respected, and there was a large company of friends following to her grave. She had been brought up on one of the neighboring farms, and each of the few times that I had seen her she professed great dissatisfaction with town life. The people lived too close together for her liking, at the Landing, and she could not get used to the constant sound of the sea. She had lived to lament three seafaring husbands, and her house was decorated with West Indian curiosities, specimens of conch shells and fine coral which they had brought home from their voyages in lumber-laden ships. Mrs Todd had told me all our neighbor's history. They had been girls together, and, to use her own phrase, had 'both seen trouble till they know the best and worst on 't.' I could see the sorrowful, large figure of Mrs Todd as I stood at the window. She made a break in the procession by walking slowly and keeping the after-part of it back. She held a handkerchief to her eyes, and I knew, with a pang of sympathy, that hers was not affected grief.

Beside her, after much difficulty, I recognized the one strange and unrelated person in all the company, an old man who had always been mysterious to me. I could see his thin, bending figure. He wore a narrow, long-tailed coat and walked with a stick, and had the same 'cant to leeward' as the wind-bent trees on the height above.

This was Captain Littlepage, whom I had seen only once or

twice before, sitting pale and old behind a closed window; never out of doors until now. Mrs Todd always shook her head gravely when I asked a question, and said that he wasn't what he had been once, and seemed to class him with her other secrets. He might have belonged with a simple which grew in a certain slug-haunted corner of the garden, whose use she could never be betrayed into telling me, though I saw her cutting the tops by moonlight once, as if it were a charm, and not a medicine, like the great fading bloodroot leaves.

I could see that she was trying to keep pace with the old captain's lighter steps. He looked like an aged grasshopper of some strange human variety. Behind this pair was a short, impatient, little person, who kept the captain's house, and gave it what Mrs Todd and others believed to be no proper sort of care. She was usually called 'that Mari' Harris' in subdued conversation between intimates, but they treated her with anxious civility when they met her face to face.

The bay-sheltered islands and the great sea beyond stretched away to the far horizon southward and eastward; the little procession in the foreground looked futile and helpless on the edge of the rocky shore. It was a glorious day early in July, with a clear, high sky; there were no clouds, there was no noise of the sea. The song sparrows sang and sang, as if with joyous knowledge of immortality, and contempt for those who could so pettily concern themselves with death. I stood watching until the funeral procession had crept round a shoulder of the slope below and disappeared from the great landscape as if it had gone into a cave.

An hour later I was busy at my work. Now and then a bee blundered in and took me for an enemy; but there was a useful stick upon the teacher's desk, and I rapped to call the bees to order as if they were unruly scholars, or waved them away from their riots over the ink, which I had bought at the Landing store, and discovered too late to be scented with bergamot, as if to refresh the labors of anxious scribes. One anxious scribe felt very dull that day; a sheep-bell tinkled near by, and called her wandering wits after it. The sentences failed to catch these lovely summer cadences. For the first

time I began to wish for a companion and for news from the
outer world, which had been, half unconsciously, forgotten.
Watching the funeral gave one a sort of pain. I began to
wonder if I ought not to have walked with the rest, instead of
hurrying away at the end of the services. Perhaps the Sunday
gown I had put on for the occasion was making this disastrous
change of feeling, but I had now made myself and my friends
remember that I did not really belong to Dunnet Landing.
  I sighed, and turned to the half-written page again.

V

CAPTAIN LITTLEPAGE

It was a long time after this; an hour was very long in that coast town where nothing stole away the shortest minute. I had lost myself completely in work, when I heard footsteps outside. There was a steep footpath between the upper and the lower road, which I climbed to shorten the way, as the children had taught me, but I believed that Mrs Todd would find it inaccessible, unless she had occasion to seek me in great haste. I wrote on, feeling like a besieged miser of time, while the footsteps came nearer, and the sheep-bell tinkled away in haste as if some one had shaken a stick in its wearer's face. Then I looked, and saw Captain Littlepage passing the nearest window; the next moment he tapped politely at the door.

'Come in, sir,' I said, rising to meet him; and he entered, bowing with much courtesy. I stepped down from the desk and offered him a chair by the window, where he seated himself at once, being sadly spent by his climb. I returned to my fixed seat behind the teacher's desk, which gave him the lower place of a scholar.

'You ought to have the place of honor, Captain Littlepage,' I said.

'A happy, rural seat of various views,'*

he quoted, as he gazed out into the sunshine and up the long wooded shore. Then he glanced at me, and looked all about him as pleased as a child.

'My quotation was from Paradise Lost: the greatest of poems, I suppose you know?' and I nodded. 'There's nothing that ranks, to my mind, with Paradise Lost; it's all lofty, all lofty,' he continued. 'Shakespeare was a great poet; he copied life, but you have to put up with a great deal of low talk.'

I now remembered that Mrs Todd had told me one day that Captain Littlepage had overset his mind with too much reading; she had also made dark reference to his having 'spells' of some unexplainable nature. I could not help wondering what errand had brought him out in search of me. There was something quite charming in his appearance: it was a face thin and delicate with refinement, but worn into appealing lines, as if he had suffered from loneliness and misapprehension. He looked, with his careful precision of dress, as if he were the object of cherishing care on the part of elderly unmarried sisters, but I knew Mari' Harris to be a very commonplace, inelegant person, who would have no such standards; it was plain that the captain was his own attentive valet. He sat looking at me expectantly. I could not help thinking that, with his queer head and length of thinness, he was made to hop along the road of life rather than to walk. The captain was very grave indeed, and I bade my inward spirit keep close to discretion.

'Poor Mrs Begg has gone,' I ventured to say. I still wore my Sunday gown by way of showing respect.

'She has gone,' said the captain,—'very easy at the last, I was informed; she slipped away as if she were glad of the opportunity.'

I thought of the Countess of Carberry* and felt that history repeated itself.

'She was one of the old stock,' continued Captain Littlepage, with touching sincerity. 'She was very much looked up to in this town, and will be missed.'

I wondered, as I looked at him, if he had sprung from a line of ministers; he had the refinement of look and air of command which are the heritage of the old ecclesiastical families of New England. But as Darwin says in his autobiography, 'there is no such king as a sea-captain; he is greater even than a king or a schoolmaster!'*

Captain Littlepage moved his chair out of the wake of the sunshine, and still sat looking at me. I began to be very eager to know upon what errand he had come.

'It may be found out some o' these days,' he said earnestly. 'We may know it all, the next step; where Mrs Begg is now, for instance. Certainty, not conjecture, is what we all desire.'

'I suppose we shall know it all some day,' said I.

'We shall know it while yet below,' insisted the captain, with a flush of impatience on his thin cheeks. 'We have not looked for truth in the right direction. I know what I speak of; those who have laughed at me little know how much reason my ideas are based upon.' He waved his hand toward the village below. 'In that handful of houses they fancy that they comprehend the universe.'

I smiled, and waited for him to go on.

'I am an old man, as you can see,' he continued, 'and I have been a shipmaster the greater part of my life,—forty-three years in all. You may not think it, but I am above eighty years of age.'

He did not look so old, and I hastened to say so.

'You must have left the sea a good many years ago, then, Captain Littlepage?' I said.

'I should have been serviceable at least five or six years more,' he answered. 'My acquaintance with certain—my experience upon a certain occasion, I might say, gave rise to prejudice. I do not mind telling you that I chanced to learn of one of the greatest discoveries that man has ever made.'

Now we were approaching dangerous ground, but a sudden sense of his sufferings at the hands of the ignorant came to my help, and I asked to hear more with all the deference I really felt. A swallow flew into the schoolhouse at this moment as if a kingbird were after it, and beat itself against the walls for a minute, and escaped again to the open air; but Captain Littlepage took no notice whatever of the flurry.

'I had a valuable cargo of general merchandise from the London docks to Fort Churchill, a station of the old company on Hudson's Bay,' said the captain earnestly. 'We were delayed in lading, and baffled by head winds and a heavy tumbling sea all the way north-about and across. Then the fog kept us off the coast; and when I made port at last, it was too late to delay in those northern waters with such a vessel and such a crew as I had. They cared for nothing, and idled me into a fit of sickness; but my first mate was a good, excellent man, with no more idea of being frozen in there until spring than I had, so we made what speed we could to get clear of Hudson's Bay and off the coast. I owned an eighth of the vessel, and he owned a sixteenth of her. She was a full-rigged ship, called the Minerva, but she was getting old and leaky. I meant it should be my last v'y'ge in her, and so it proved. She had been an excellent vessel in her day. Of the cowards aboard her I can't say so much.'

'Then you were wrecked?' I asked, as he made a long pause.

'I wa'n't caught astern o' the lighter by any fault of mine,' said the captain gloomily. 'We left Fort Churchill and run out into the Bay with a light pair o' heels; but I had been vexed to death with their red-tape rigging at the company's office, and chilled with stayin' on deck an' tryin' to hurry up things, and when we were well out o' sight o' land, headin' for Hudson's Straits, I had a bad turn o' some sort o' fever, and had to stay below. The days were getting short, and we made good runs, all well on board but me, and the crew done their work by dint of hard driving.'

I began to find this unexpected narrative a little dull. Captain Littlepage spoke with a kind of slow correctness that lacked the longshore high flavor to which I had grown used; but I listened respectfully while he explained the winds having become contrary, and talked on in a dreary sort of way about his voyage, the bad weather, and the disadvantages he was under in the lightness of his ship, which bounced about like a chip in a bucket, and would not answer the rudder or properly respond to the most careful setting of sails.

'So there we were blowin' along anyways,' he complained; but looking at me at this moment, and seeing that my thoughts were unkindly wandering, he ceased to speak.

'It was a hard life at sea in those days, I am sure,' said I, with redoubled interest.

'It was a dog's life,' said the poor old gentleman, quite reassured, 'but it made men of those who followed it. I see a change for the worse even in our own town here; full of loafers now, small and poor as 't is, who once would have followed the sea, every lazy soul of 'em. There is no occupation so fit for just that class o' men who never get beyond the fo'cas'le. I view it, in addition, that a community narrows down and grows dreadful ignorant when it is shut up to its own affairs, and gets no knowledge of the outside world except from a cheap, unprincipled newspaper. In the old days, a good part o' the best men here knew a hundred ports and something of the way folks lived in them. They saw the world for themselves, and like 's not their wives and children saw it with them. They may not have had the best of knowledge to carry with 'em sight-seein', but they were some acquainted with foreign lands an' their laws, an' could see outside the battle for town clerk here in Dunnet; they got some sense o' proportion. Yes, they lived more dignified, and their houses were better within an' without. Shipping's a terrible loss to this part o' New England from a social point o' view, ma'am.'

'I have thought of that myself,' I returned, with my interest quite awakened. 'It accounts for the change in a great many things,—the sad disappearance of sea-captains,—doesn't it?'

'A shipmaster was apt to get the habit of reading,' said my companion, brightening still more, and taking on a most touching air of unreserve. 'A captain is not expected to be familiar with his crew, and for company's sake in dull days and nights he turns to his book. Most of us old shipmasters came to know 'most everything about something; one would take to readin' on farming topics, and some were great on medicine,—but Lord help their poor crews!—or some were all for history, and now and then there'd be one like me that gave his time to the poets. I was well acquainted with a shipmaster that was all for bees an' bee-keepin'; and if you met him in port and went aboard, he'd sit and talk a terrible while about their havin' so much information, and the money that could be made out of keepin' 'em. He was one of the smartest captains that ever sailed the seas, but they used to call the Newcastle, a great bark he commanded for many years, Tuttle's beehive. There was old Cap'n Jameson: he had notions of Solomon's Temple,* and made a very handsome little model of the same, right from the Scripture measurements, same's other sailors make little ships and design new tricks of rigging and all that. No, there's nothing to take the place of shipping in a place like ours. These bicycles offend me dreadfully; they don't afford no real opportunities of experience such as a man gained on a voyage. No: when folks left home in the old days they left it to some purpose, and when they got home they stayed there and had some pride in it. There's no large-minded way of thinking now: the worst have got to be best and rule everything; we're all turned upside down and going back year by year.'

'Oh no, Captain Littlepage, I hope not,' said I, trying to soothe his feelings.

There was a silence in the schoolhouse, but we could hear the noise of the water on a beach below. It sounded like the strange warning wave that gives notice of the turn of the tide. A late golden robin, with the most joyful and eager of voices, was singing close by in a thicket of wild roses.

VI

THE WAITING PLACE

'How did you manage with the rest of that rough voyage on the Minerva?' I asked.

'I shall be glad to explain to you,' said Captain Littlepage, forgetting his grievances for the moment. 'If I had a map at hand I could explain better. We were driven to and fro 'way up toward what we used to call Parry's Discoveries,* and lost our bearings. It was thick and foggy, and at last I lost my ship; she drove on a rock, and we managed to get ashore on what I took to be a barren island, the few of us that were left alive. When she first struck, the sea was somewhat calmer than it had been, and most of the crew, against orders, manned the long-boat and put off in a hurry, and were never heard of more. Our own boat upset, but the carpenter kept himself and me above water, and we drifted in. I had no strength to call upon after my recent fever, and laid down to die; but he found the tracks of a man and dog the second day, and got along the shore to one of those far missionary stations that the Moravians support. They were very poor themselves, and in distress; 't was a useless place. There were but few Esquimaux left in that region. There we remained for some time, and I became acquainted with strange events.'

The captain lifted his head and gave me a questioning glance. I could not help noticing that the dulled look in his eyes had gone, and there was instead a clear intentness that made them seem dark and piercing.

'There was a supply ship expected, and the pastor, an excellent Christian man, made no doubt that we should get passage in her. He was hoping that orders would come to break up the station; but everything was uncertain, and we got on the best we could for a while. We fished, and helped the people in other ways; there was no other way of paying our debts. I was taken to the pastor's house until I got better; but they were crowded, and I felt myself in the way, and made excuse to join with an old seaman, a Scotchman, who had built him a warm cabin, and had room in it for another. He was looked upon with regard, and had stood by the pastor in some troubles with the people. He had been on one of those English exploring parties that found one end of the road to the north pole, but never could find the other. We lived like dogs in a kennel, or so you'd thought if you had seen the hut from the outside; but the main thing was to keep warm; there were piles of birdskins to lie on, and he'd made him a good bunk, and there was another for me. 'T was dreadful dreary waitin' there; we begun to think the supply steamer was lost, and my poor ship broke up and strewed herself all along the shore. We got to watching on the headlands; my men and me knew the people were short of supplies and had to pinch themselves. It ought to read in the Bible, "Man cannot live by fish alone,"* if they'd told the truth of things; 't aint bread that wears the worst on you! First part of the time, old Gaffett, that I lived with, seemed speechless, and I didn't know what to make of him, nor he of me, I dare say; but as we got acquainted, I found he'd been through more disasters than I had, and had troubles that wa'n't going to let him live a great while. It used to ease his mind to talk to an understanding person, so we used to sit and talk together all day, if it rained or blew so that we couldn't get out. I'd got a bad blow on the back of my head at the time we came ashore, and it pained me at times, and my strength was broken, anyway; I've never been so able since.'

Captain Littlepage fell into a reverie.

'Then I had the good of my reading,' he explained presently. 'I had no books; the pastor spoke but little English, and all his books were foreign; but I used to say over all I could remember. The old poets little knew what comfort they could be to a man. I was well acquainted with the works of Milton, but up there it did seem to me as if Shakespeare was the king; he has his sea terms very accurate, and some beautiful passages were calming to the mind. I could say them over until I shed tears; there was nothing beautiful to me in that place but the stars above and those passages of verse.

'Gaffett was always brooding and brooding, and talking to

himself; he was afraid he should never get away, and it preyed upon his mind. He thought when I got home I could interest the scientific men in his discovery: but they're all taken up with their own notions; some didn't even take pains to answer the letters I wrote. You observe that I said this crippled man Gaffett had been shipped on a voyage of discovery. I now tell you that the ship was lost on its return, and only Gaffett and two officers were saved off the Greenland coast, and he had knowledge later that those men never got back to England; the brig they shipped on was run down in the night. So no other living soul had the facts, and he gave them to me. There is a strange sort of a country 'way up north beyond the ice, and strange folks living in it. Gaffett believed it was the next world to this.'

'What do you mean, Captain Littlepage?' I exclaimed. The old man was bending forward and whispering; he looked over his shoulder before he spoke the last sentence.

'To hear old Gaffett tell about it was something awful,' he said, going on with his story quite steadily after the moment of excitement had passed. ' 'T was first a tale of dogs and sledges, and cold and wind and snow. Then they begun to find the ice grow rotten; they had been frozen in, and got into a current flowing north, far up beyond Fox Channel,* and they took to their boats when the ship got crushed, and this warm current took them out of sight of the ice, and into a great open sea; and they still followed it due north, just the very way they had planned to go. Then they struck a coast that wasn't laid down or charted, but the cliffs were such that no boat could land until they found a bay and struck across under sail to the other side where the shore looked lower; they were scant of provisions and out of water, but they got sight of something that looked like a great town. "For God's sake, Gaffett!" said I, the first time he told me. "You don't mean a town two degrees farther north than ships had ever been?" for he'd got their course marked on an old chart that he'd pieced out at the top; but he insisted upon it, and told it over and over again, to be sure I had it straight to carry to those who would be interested. There was no snow and ice, he said, after they had sailed some days with that warm current, which seemed to come right from under the ice that they'd been pinched up in and had been crossing on foot for weeks.'

'But what about the town?' I asked. 'Did they get to the town?'

'They did,' said the captain, 'and found inhabitants; 't was an awful condition of things. It appeared, as near as Gaffett could express it, like a place where there was neither living nor dead. They could see the place when they were approaching it by sea pretty near like any town, and thick with habitations; but all at once they lost sight of it altogether, and when they got close inshore they could see the shapes of folks, but they never could get near them,—all blowing gray figures that would pass along alone, or sometimes gathered in companies as if they were watching. The men were frightened at first, but the shapes never came near them,—it was as if they blew back; and at last they all got bold and went ashore, and found birds' eggs and sea fowl, like any wild northern spot where creatures were tame and folks had never been, and there was good water. Gaffett said that he and another man came near one o' the fog-shaped men that was going along slow with the look of a pack on his back, among the rocks, an' they chased him; but, Lord! he flittered away out o' sight like a leaf the wind takes with it, or a piece of cobweb. They would make as if they talked together, but there was no sound of voices, and "they acted as if they didn't see us, but only felt us coming towards them," says Gaffett one day, trying to tell the particulars. They couldn't see the town when they were ashore. One day the captain and the doctor were gone till night up across the high land where the town had seemed to be, and they came back at night beat out and white as ashes, and wrote and wrote all next day in their notebooks, and whispered together full of excitement, and they were sharp-spoken with the men when they offered to ask any questions.

'Then there came a day,' said Captain Littlepage, leaning toward me with a strange look in his eyes, and whispering quickly. 'The men all swore they wouldn't stay any longer; the man on watch early in the morning gave the alarm, and they all put off in the boat and got a little way out to sea. Those

folks, or whatever they were, come about 'em like bats; all at
once they raised incessant armies,* and come as if to drive
'em back to sea. They stood thick at the edge o' the water like
the ridges o' grim war; no thought o' flight, none of retreat.
Sometimes a standing fight, then soaring on main wing tor-
mented all the air. And when they'd got the boat out o' reach
o' danger, Gaffett said they looked back, and there was the
town again, standing up just as they'd seen it first, comin' on
the coast. Say what you might, they all believed 't was a kind of
waiting-place between this world an' the next.'

The captain had sprung to his feet in his excitement, and
made excited gestures, but he still whispered huskily.

'Sit down, sir,' I said as quietly as I could, and he sank into
his chair quite spent.

'Gaffett thought the officers were hurrying home to report
and to fit out a new expedition when they were all lost. At the
time, the men got orders not to talk over what they had seen,'
the old man explained presently in a more natural tone.

'Weren't they all starving, and wasn't it a mirage or something of that sort?' I ventured to ask. But he looked at me
blankly.

'Gaffett had got so that his mind ran on nothing else,' he
went on. 'The ship's surgeon let fall an opinion to the captain, one day, that 't was some condition o' the light and the
magnetic currents that let them see those folks. 'T wa'n't a
right-feeling part of the world, anyway; they had to battle with
the compass to make it serve, an' everything seemed to go
wrong. Gaffett had worked it out in his own mind that they
was all common ghosts, but the conditions were unusual
favorable for seeing them. He was always talking about the
Ge'graphical Society,* but he never took proper steps, as I
view it now, and stayed right there at the mission. He was a
good deal crippled, and thought they'd confine him in some
jail of a hospital. He said he was waiting to find the right men
to tell, somebody bound north. Once in a while they stopped
there to leave a mail or something. He was set in his notions,
and let two or three proper explorin' expeditions go by him
because he didn't like their looks; but when I was there he
had got restless, fearin' he might be taken away or something.

He had all his directions written out straight as a string to give the right ones. I wanted him to trust 'em to me, so I might have something to show, but he wouldn't. I suppose he's dead now. I wrote to him, an' I done all I could. 'T will be a great exploit some o' these days.'

I assented absent-mindedly, thinking more just then of my companion's alert, determined look and the seafaring, ready aspect that had come to his face; but at this moment there fell a sudden change, and the old, pathetic, scholarly look returned. Behind me hung a map of North America, and I saw, as I turned a little, that his eyes were fixed upon the northernmost regions and their careful recent outlines* with a look of bewilderment.

VII

THE OUTER ISLAND

GAFFETT with his good bunk and the bird-skins, the story of the wreck of the Minerva, the human-shaped creatures of fog and cobweb, the great words of Milton with which he described their onslaught upon the crew, all this moving tale had such an air of truth that I could not argue with Captain Littlepage. The old man looked away from the map as if it had vaguely troubled him, and regarded me appealingly.

'We were just speaking of'—and he stopped. I saw that he had suddenly forgotten his subject.

'There were a great many persons at the funeral,' I hastened to say.

'Oh yes,' the captain answered, with satisfaction. 'All showed respect who could. The sad circumstances had for a moment slipped my mind. Yes, Mrs Begg will be very much missed. She was a capital manager for her husband when he was at sea. Oh yes, shipping is a very great loss.'* And he sighed heavily. 'There was hardly a man of any standing who didn't interest himself in some way in navigation. It always gave credit to a town. I call it low-water mark now here in Dunnet.'

He rose with dignity to take leave, and asked me to stop at his house some day, when he would show me some outlandish things that he had brought home from sea. I was familiar with the subject of the decadence of shipping interests in all its affecting branches, having been already some time in Dunnet, and I felt sure that Captain Littlepage's mind had now returned to a safe level.

As we came down the hill toward the village our ways divided, and when I had seen the old captain well started on a smooth piece of sidewalk which would lead him to his own door, we parted, the best of friends. 'Step in some afternoon,' he said, as affectionately as if I were a fellow-shipmaster wrecked on the lee shore of age like himself. I turned toward home, and presently met Mrs Todd coming toward me with an anxious expression.

'I see you sleevin'* the old gentleman down the hill,' she suggested.

'Yes. I've had a very interesting afternoon with him,' I answered; and her face brightened.

'Oh, then he's all right. I was afraid 't was one o' his flighty spells, an' Mari' Harris wouldn't't'—

'Yes,' I returned, smiling, 'he has been telling me some old stories, but we talked about Mrs Begg and the funeral beside, and Paradise Lost.'

'I expect he got tellin' of you some o' his great narratives,' she answered, looking at me shrewdly. 'Funerals always sets him goin'. Some o' them tales hangs together toler'ble well,' she added, with a sharper look than before. 'An' he's been a great reader all his seafarin' days. Some thinks he overdid, and affected his head, but for a man o' his years he's amazin' now when he's at his best. Oh, he used to be a beautiful man!'

We were standing where there was a fine view of the harbor and its long stretches of shore all covered by the great army of the pointed firs, darkly cloaked and standing as if they waited to embark. As we looked far seaward among the outer islands, the trees seemed to march seaward still, going steadily over the heights and down to the water's edge.

It had been growing gray and cloudy, like the first evening of autumn, and a shadow had fallen on the darkening shore. Suddenly, as we looked, a gleam of golden sunshine struck the outer islands, and one of them shone out clear in the light, and revealed itself in a compelling way to our eyes. Mrs Todd was looking off across the bay with a face full of affection and interest. The sunburst upon that outermost island made it seem like a sudden revelation of the world beyond this which some believe to be so near.

'That's where mother lives,' said Mrs Todd. 'Can't we see it plain? I was brought up out there on Green Island. I know every rock an' bush on it.'

'Your mother!' I exclaimed, with great interest.

'Yes, dear, cert'in; I've got her yet, old 's I be. She's one of them spry, light-footed little women; always was, an' light-hearted, too,' answered Mrs Todd, with satisfaction. 'She's seen all the trouble folks can see, without it's her last sickness; an' she's got a word of courage for everybody. Life ain't spoilt her a mite. She's eighty-six an' I'm sixty-seven, and I've seen the time I've felt a good sight the oldest. "Land sakes alive!" says she, last time I was out to see her. "How you do lurch about steppin' into a bo't!" I laughed so I liked to have gone right over into the water; an' we pushed off, an' left her laughin' there on the shore.'

The light had faded as we watched. Mrs Todd had mounted a gray rock, and stood there grand and architectural, like a *caryatide.* Presently she stepped down, and we continued our way homeward.

'You an' me, we'll take a bo't an' go out some day and see mother,' she promised me. ''T would please her very much, an' there's one or two sca'ce herbs grows better on the island than anywheres else. I ain't seen their like nowheres here on the main.'

'Now I'm goin' right down to get us each a mug o' my beer,' she announced as we entered the house, 'an' I believe I'll sneak in a little mite o' camomile. Goin' to the funeral an' all, I feel to have had a very wearin' afternoon.'

I heard her going down into the cool little cellar, and then there was considerable delay. When she returned, mug in hand, I noticed the taste of camomile, in spite of my protest; but its flavor was disguised by some other herb that I did not know, and she stood over me until I drank it all and said that I liked it.

'I don't give that to everybody,' said Mrs Todd kindly; and I felt for a moment as if it were part of a spell and incantation, and as if my enchantress would now begin to look like the cobweb shapes of the arctic town. Nothing happened but a quiet evening and some delightful plans that we made about going to Green Island, and on the morrow there was the clear sunshine and blue sky of another day.

VIII

GREEN ISLAND

ONE morning, very early, I heard Mrs Todd in the garden outside my window. By the unusual loudness of her remarks to a passer-by, and the notes of a familiar hymn which she sang as she worked among the herbs, and which came as if directed purposely to the sleepy ears of my consciousness, I knew that she wished I would wake up and come and speak to her.

In a few minutes she responded to a morning voice from behind the blinds. 'I expect you're goin' up to your schoolhouse to pass all this pleasant day; yes, I expect you're goin' to be dreadful busy,' she said despairingly.

'Perhaps not,' said I. 'Why, what's going to be the matter with you, Mrs Todd?' For I supposed that she was tempted by the fine weather to take one of her favorite expeditions along the shore pastures to gather herbs and simples, and would like to have me keep the house.

'No, I don't want to go nowhere by land,' she answered gayly,—'no, not by land; but I don't know 's we shall have a better day all the rest of the summer to go out to Green Island an' see mother. I waked up early thinkin' of her. The wind's light northeast,—'t will take us right straight out; an' this time o' year it's liable* to change round southwest an' fetch us home pretty, 'long late in the afternoon. Yes, it's goin' to be a good day.'

'Speak to the captain and the Bowden boy, if you see anybody going by toward the landing,' said I. 'We'll take the big boat.'

'Oh, my sakes! now you let me do things my way,' said Mrs Todd scornfully. 'No, dear, we won't take no big bo't. I'll just git a handy dory, an' Johnny Bowden an' me, we'll man her ourselves. I don't want no abler bo't than a good dory, an' a nice light breeze ain't goin' to make no sea; an' Johnny's my cousin's son,—mother'll like to have him come; an' he'll be down to the herrin' weirs all the time we're there, anyway; we

don't want to carry no men folks havin' to be considered
every minute an' takin' up all our time. No, you let me do;
we'll just slip out an' see mother by ourselves. I guess what
breakfast you'll want's about ready now.'

I had become well acquainted with Mrs Todd as landlady,
herb-gatherer, and rustic philosopher; we had been discreet
fellow-passengers once or twice when I had sailed up the coast
to a larger town than Dunnet Landing to do some shopping;
but I was yet to become acquainted with her as a mariner. An
hour later we pushed off from the landing in the desired
dory. The tide was just on the turn, beginning to fall, and
several friends and acquaintances stood along the side of the
dilapidated wharf and cheered us by their words and evident
interest. Johnny Bowden and I were both rowing in haste to
get out where we could catch the breeze and put up the small
sail which lay clumsily furled along the gunwale. Mrs Todd sat
aft, a stern and unbending law-giver.

'You better let her drift; we'll get there 'bout as quick; the
tide'll take her right out from under these old buildin's;
there's plenty wind outside.'

'Your bo't ain't trimmed proper, Mis' Todd!' exclaimed a
voice from shore. 'You're lo'ded so the bo't'll drag; you can't
git her before the wind, ma'am. You set 'midships, Mis' Todd,
an' let the boy hold the sheet 'n' steer after he gits the sail up;
you won't never git out to Green Island that way. She's lo'ded
bad, your bo't is,—she's heavy behind 's she is now!'

Mrs Todd turned with some difficulty and regarded the
anxious adviser, my right oar flew out of water, and we
seemed about to capsize. 'That you, Asa? Good-mornin',' she
said politely. 'I al'ays liked the starn seat best. When'd you git
back from up country?'

This allusion to Asa's origin was not lost upon the rest of
the company. We were some little distance from shore, but we
could hear a chuckle of laughter, and Asa, a person who was
too ready with his criticism and advice on every possible
subject, turned and walked indignantly away.

When we caught the wind we were soon on our seaward
course, and only stopped to underrun a trawl, for the floats of
which Mrs Todd looked earnestly, explaining that her mother

might not be prepared for three extra to dinner; it was her brother's trawl, and she meant to just run her eye along for the right sort of a little haddock. I leaned over the boat's side with great interest and excitement, while she skillfully handled the long line of hooks, and made scornful remarks upon worthless, bait-consuming creatures of the sea as she reviewed them and left them on the trawl or shook them off into the waves. At last we came to what she pronounced a proper haddock, and having taken him on board and ended his life resolutely, we went our way.

As we sailed along I listened to an increasingly delightful commentary upon the islands, some of them barren rocks, or at best giving sparse pasturage for sheep in the early summer. On one of these an eager little flock ran to the water's edge and bleated at us so affectingly that I would willingly have stopped; but Mrs Todd steered away from the rocks, and scolded at the sheep's mean owner, an acquaintance of hers, who grudged the little salt and still less care which the patient creatures needed. The hot midsummer sun makes prisons of these small islands that are a paradise in early June, with their cool springs and short thick-growing grass. On a larger island, farther out to sea, my entertaining companion showed me with glee the small houses of two farmers who shared the island between them, and declared that for three generations the people had not spoken to each other even in times of sickness or death or birth. 'When the news come that the war was over, one of 'em knew it a week, and never stepped across his wall to tell the others,' she said. 'There, they enjoy it: they've got to have somethin' to interest 'em in such a place; 't is a good deal more tryin' to be tied to folks you don't like than 't is to be alone, Each of 'em tells the neighbors their wrongs; plenty likes to hear and tell again; them as fetch a bone'll carry one,* an' so they keep the fight a-goin'. I must say I like variety myself; some folks washes Monday an' irons Tuesday the whole year round, even if the circus is goin' by!'

A long time before we landed at Green Island we could see the small white house, standing high like a beacon, where Mrs Todd was born and where her mother lived, on a green slope above the water, with dark spruce woods still higher. There were crops in the fields, which we presently distinguished from one another. Mrs Todd examined them while we were still far at sea. 'Mother's late potatoes looks backward; ain't had rain enough so far,' she pronounced her opinion. 'They look weedier than what they call Front Street down to Cowper Centre. I expect brother William is so occupied with his herrin' weirs an' servin' out bait to the schooners that he don't think once a day of the land,'

'What's the flag for, up above the spruces there behind the house?' I inquired, with eagerness.

'Oh, that's the sign for herrin',' she explained kindly, while Johnny Bowden regarded me with contemptuous surprise. 'When they get enough for schooners they raise that flag; an' when 't is a poor catch in the weir pocket they just fly a little signal down by the shore, an' then the small bo'ts comes and get enough an' over for their trawls. There, look! there she is: mother sees us; she's wavin' somethin' out o' the fore door! She'll be to the landin'-place quick 's we are.'

I looked, and could see a tiny flutter in the doorway, but a quicker signal had made its way from the heart on shore to the heart on the sea.

'How do you suppose she knows it's me?' said Mrs Todd, with a tender smile on her broad face. 'There, you never get over bein' a child long 's you have a mother to go to. Look at the chimney, now; she's gone right in an' brightened up the fire. Well, there, I'm glad mother's well; you'll enjoy seein' her very much.'

Mrs Todd leaned back into her proper position, and the boat trimmed again. She took a firmer grasp of the sheet, and gave an impatient look up at the gaff and the leech of the little sail, and twitched the sheet as if she urged the wind like a horse. There came at once a fresh gust, and we seemed to have doubled our speed. Soon we were near enough to see a tiny figure with handkerchiefed head come down across the field and stand waiting for us at the cove above a curve of pebble beach.

Presently the dory grated on the pebbles, and Johnny Bowden, who had been kept in abeyance during the voyage, sprang out and used manful exertions to haul us up with the next wave, so that Mrs Todd could make a dry landing.

'You done that very well,' she said, mounting to her feet, and coming ashore somewhat stiffly, but with great dignity, refusing our outstretched hands, and returning to possess herself of a bag which had lain at her feet.

'Well, mother, here I be!' she announced with indifference; but they stood and beamed in each other's faces.

'Lookin' pretty well for an old lady, ain't she?' said Mrs Todd's mother, turning away from her daughter to speak to me. She was a delightful little person herself, with bright eyes and an affectionate air of expectation like a child on a holiday. You felt as if Mrs Blackett were an old and dear friend before you let go her cordial hand. We all started together up the hill.

'Now don't you haste too fast, mother,' said Mrs Todd warningly; ''t is a far reach o' risin' ground to the fore door, and you won't set an' get your breath when you're once there, but go trotting about. Now don't you go a mite faster than we proceed with this bag an' basket. Johnny, there, 'll fetch up the haddock. I just made one stop to underrun William's trawl till I come to jes' such a fish 's I thought you'd want to make one o' your nice chowders of. I've brought an onion with me that was layin' about on the window-sill at home.'

'That's just what I was wantin',' said the hostess. 'I give a sigh when you spoke o' chowder, knowin' my onions was out. William forgot to replenish us last time he was to the Landin'. Don't you haste so yourself, Almiry, up this risin' ground. I hear you commencin' to wheeze a'ready.'

This mild revenge seemed to afford great pleasure to both giver and receiver. They laughed a little, and looked at each other affectionately, and then at me. Mrs Todd considerately paused, and faced about to regard the wide sea view. I was glad to stop, being more out of breath than either of my companions, and I prolonged the halt by asking the names of the neighboring islands. There was a fine breeze blowing, which we felt more there on the high land than when we were running before it in the dory.

'Why, this ain't that kitten I saw when I was out last, the one that I said didn't appear likely?' exclaimed Mrs Todd as we went our way.

'That's the one, Almiry,' said her mother. 'She always had a likely look to me, an' she's right after her business. I never see such a mouser for one of her age. If 't wan't for William, I never should have housed that other dronin' old thing so long; but he sets by her on account of her havin' a bob tail. I don't deem it advisable to maintain cats just on account of their havin' bob tails; they're like all other curiosities, good for them that wants to see 'em twice. This kitten catches mice for both, an' keeps me respectable as I ain't been for a year. She's a real understandin' little help, this kitten is. I picked her from among five Miss Augusta Pennell had over to Burnt Island,' said the old woman, trudging along with the kitten close at her skirts. 'Augusta, she says to me, "Why, Mis' Blackett, you've took the homeliest;" an' says I, "I've got the smartest; I 'm satisfied."'

'I'd trust nobody sooner 'n you to pick out a kitten, mother,' said the daughter handsomely, and we went on in peace and harmony.

The house was just before us now, on a green level that looked as if a huge hand had scooped it out of the long green field we had been ascending. A little way above, the dark spruce woods began to climb the top of the hill and cover the seaward slopes of the island. There was just room for the small farm and the forest; we looked down at the fish-house and its rough sheds, and the weirs stretching far out into the water. As we looked upward, the tops of the firs came sharp against the blue sky. There was a great stretch of rough pasture-land round the shoulder of the island to the eastward, and here were all the thick-scattered gray rocks that kept their places, and the gray backs of many sheep that forever wandered and fed on the thin sweet pasturage that fringed the ledges and made soft hollows and strips of green turf like growing velvet. I could see the rich green of bayberry bushes here and there, where the rocks made room. The air was very sweet; one could not help wishing to be a citizen of such a complete and tiny continent and home of fisherfolk.

The house was broad and clean, with a roof that looked heavy on its low walls. It was one of the houses that seem firm-rooted in the ground, as if they were two-thirds below the surface, like icebergs. The front door stood hospitably open in expectation of company, and an orderly vine grew at each side; but our path led to the kitchen door at the house-end, and there grew a mass of gay flowers and greenery, as if they had been swept together by some diligent garden broom into a tangled heap: there were portulacas all along under the lower step and straggling off into the grass, and clustering mallows that crept as near as they dared, like poor relations. I saw the bright eyes and brainless little heads of two half-grown chickens who were snuggled down among the mallows as if they had been chased away from the door more than once, and expected to be again.

'It seems kind o' formal comin' in this way,' said Mrs Todd impulsively, as we passed the flowers and came to the front doorstep; but she was mindful of the proprieties, and walked before us into the best room on the left.

'Why, mother, if you haven't gone an' turned the carpet!' she exclaimed, with something in her voice that spoke of awe and admiration. 'When'd you get to it? I s'pose Mis' Addicks come over an' helped you, from White Island Landing?'

'No, she didn't,' answered the old woman, standing proudly erect, and making the most of a great moment. 'I done it all myself with William's help. He had a spare day, an' took right holt with me; an' 't was all well beat on the grass, an' turned, an' put down again afore we went to bed. I ripped an' sewed over two o' them long breadths.* I ain't had such a good night's sleep for two years.'

'There, what do you think o' havin' such a mother as that for eighty-six year old?' said Mrs Todd, standing before us like a large figure of Victory.*

As for the mother, she took on a sudden look of youth; you felt as if she promised a great future, and was beginning, not ending, her summers and their happy toils.

'My, my!' exclaimed Mrs Todd. 'I couldn't ha' done it myself, I've got to own it.'

'I was much pleased to have it off my mind,' said Mrs Blackett, humbly; 'the more so because along at the first of the next week I wasn't very well. I suppose it may have been the change of weather.'

Mrs Todd could not resist a significant glance at me, but, with charming sympathy, she forbore to point the lesson or to connect this illness with its apparent cause. She loomed larger than ever in the little old-fashioned best room, with its few pieces of good furniture and pictures of national interest. The green paper curtains were stamped with conventional landscapes of a foreign order,—castles on inaccessible crags, and lovely lakes with steep wooded shores; under-foot the treasured carpet was covered thick with home-made rugs. There were empty glass lamps and crystallized bouquets* of grass and some fine shells on the narrow mantelpiece.

'I was married in this room,' said Mrs Todd unexpectedly; and I heard her give a sigh after she had spoken, as if she could not help the touch of regret that would forever come with all her thoughts of happiness.

'We stood right there between the windows,' she added, 'and the minister stood here. William wouldn't come in. He was always odd about seein' folks, just 's he is now. I run to meet 'em from a child, an' William, he'd take an' run away.'

'I've been the gainer,' said the old mother cheerfully. 'William has been son an' daughter both since you was married off the island. He's been 'most too satisfied to stop at home 'long o' his old mother, but I always tell 'em I'm the gainer.'

We were all moving toward the kitchen as if by common instinct. The best room was too suggestive of serious occasions, and the shades were all pulled down to shut out the summer light and air. It was indeed a tribute to Society to find a room set apart for her behests out there on so apparently neighborless and remote an island. Afternoon visits and evening festivals must be few in such a bleak situation at certain seasons of the year, but Mrs Blackett was of those who do not live to themselves, and who have long since passed the line that divides mere self-concern from a valued share in whatever Society can give and take. There were those of her neighbors who never had taken the trouble to furnish a best room, but Mrs Blackett was one who knew the uses of a parlor.

'Yes, do come right out into the old kitchen; I shan't make any stranger of you,' she invited us pleasantly, after we had been properly received in the room appointed to formality. 'I expect Almiry, here, 'll be driftin' out 'mongst the pasture-weeds quick 's she can find a good excuse. 'T is hot now. You'd better content yourselves till you get nice an' rested, an' 'long after dinner the sea-breeze'll spring up, an' then you can take your walks, an' go up an' see the prospect from the big ledge. Almiry'll want to show off everything there is. Then I'll get you a good cup o' tea before you start to go home. The days are plenty long now.'

While we were talking in the best room the selected fish had been mysteriously brought up from the shore, and lay all cleaned and ready in an earthen crock on the table.

'I think William might have just stopped an' said a word,' remarked Mrs Todd, pouting with high affront as she caught sight of it. 'He's friendly enough when he comes ashore, an' was remarkable social the last time, for him.'

'He ain't disposed to be very social with the ladies,' explained William's mother, with a delightful glance at me, as if she counted upon my friendship and tolerance. 'He's very particular, and he's all in his old fishin'-clothes to-day. He'll want me to tell him everything you said and done, after you've gone. William has very deep affections. He'll want to see you, Almiry. Yes, I guess he'll be in by an' by.'

'I'll search for him by 'n' by, if he don't,' proclaimed Mrs Todd, with an air of unalterable resolution. 'I know all of his burrows down 'long the shore. I'll catch him by hand 'fore he knows it. I've got some business with William, anyway. I brought forty-two cents with me that was due him for them last lobsters he brought in.'

'You can leave it with me,' suggested the little old mother, who was already stepping about among her pots and pans in the pantry, and preparing to make the chowder.

I became possessed of a sudden unwonted curiosity in regard to William, and felt that half the pleasure of my visit would be lost if I could not make his interesting acquaintance.

IX

WILLIAM

MRS TODD had taken the onion out of her basket and laid it down upon the kitchen table. 'There's Johnny Bowden come with us, you know,' she reminded her mother. 'He'll be hungry enough to eat his size.'

'I've got new doughnuts, dear,' said the little old lady. 'You don't often catch William 'n' me out o' provisions. I expect you might have chose a somewhat larger fish, but I'll try an' make it do. I shall have to have a few extra potatoes, but there's a field full out there, an' the hoe's leanin' against the well-house, in 'mongst the climbin'-beans.' She smiled, and gave her daughter a commanding nod.

'Land sakes alive! Le''s blow the horn for William,' insisted Mrs Todd, with some excitement. 'He needn't break his spirit so far 's to come in. He'll know you need him for something particular, an' then we can call to him as he comes up the path. I won't put him to no pain.'

Mrs Blackett's old face, for the first time, wore a look of trouble, and I found it necessary to counteract the teasing spirit of Almira. It was too pleasant to stay indoors altogether, even in such rewarding companionship; besides, I might meet William; and, straying out presently, I found the hoe by the well-house and an old splint basket at the woodshed door, and also found my way down to the field where there was a great square patch of rough, weedy potato-tops and tall ragweed. One corner was already dug, and I chose a fat-looking hill where the tops were well withered. There is all the pleasure that one can have in gold-digging in finding one's hopes satisfied in the riches of a good hill of potatoes. I longed to go on; but it did not seem frugal to dig any longer after my basket was full, and at last I took my hoe by the middle and lifted the basket to go back up the hill. I was sure that Mrs Blackett must be waiting impatiently to slice the potatoes into the chowder, layer after layer, with the fish.

'You let me take holt o' that basket, ma'am,' said a pleasant, anxious voice behind me.

I turned, startled in the silence of the wide field, and saw an elderly man, bent in the shoulders as fishermen often are, gray-headed and clean-shaven, and with a timid air. It was William. He looked just like his mother, and I had been imagining that he was large and stout like his sister, Almira Todd; and, strange to say, my fancy had led me to picture him not far from thirty and a little loutish. It was necessary instead to pay William the respect due to age.

I accustomed myself to plain facts on the instant, and we said good-morning like old friends. The basket was really heavy, and I put the hoe through its handle and offered him one end; then we moved easily toward the house together, speaking of the fine weather and of mackerel which were reported to be striking in all about the bay. William had been out since three o'clock, and had taken an extra fare of fish. I could feel that Mrs Todd's eyes were upon us as we approached the house, and although I fell behind in the narrow path, and let William take the basket alone and precede me at some little distance the rest of the way, I could plainly hear her greet him.

'Got round to comin' in, didn't you?' she inquired, with amusement. 'Well, now, that's clever. Didn't know 's I should see you to-day, William, an' I wanted to settle an account.'

I felt somewhat disturbed and responsible, but when I joined them they were on most simple and friendly terms. It became evident that, with William, it was the first step that cost, and that, having once joined in social interests, he was able to pursue them with more or less pleasure. He was about sixty, and not young-looking for his years, yet so undying is the spirit of youth, and bashfulness has such a power of survival, that I felt all the time as if one must try to make the occasion easy for some one who was young and new to the affairs of social life. He asked politely if I would like to go up to the great ledge while dinner was getting ready; so, not without a deep sense of pleasure, and a delighted look of surprise from the two hostesses, we started, William and I, as if both of us felt much younger than we looked. Such was the innocence and simplicity of the moment that when I heard Mrs Todd laughing behind us in the kitchen I laughed too, but William did not even blush. I think he was a little deaf, and he stepped along before me most businesslike and intent upon his errand.

We went from the upper edge of the field above the house into a smooth, brown path among the dark spruces. The hot sun brought out the fragrance of the pitchy bark, and the shade was pleasant as we climbed the hill. William stopped once or twice to show me a great wasps'-nest close by, or some fishhawks'-nests below in a bit of swamp. He picked a few sprigs of late-blooming linnæa as we came out upon an open bit of pasture at the top of the island, and gave them to me without speaking, but he knew as well as I that one could not say half he wished about linnæa. Through this piece of rough pasture ran a huge shape of stone like the great backbone of an enormous creature. At the end, near the woods, we could climb up on it and walk along to the highest point; there above the circle of pointed firs we could look down over all the island, and could see the ocean that circled this and a hundred other bits of island-ground, the mainland shore and all the far horizons. It gave a sudden sense of space, for nothing stopped the eye or hedged one in,—that sense of liberty in space and time which great prospects always give.

'There ain't no such view in the world, I expect,' said William proudly, and I hastened to speak my heartfelt tribute of praise; it was impossible not to feel as if an untraveled boy had spoken, and yet one loved to have him value his native heath.

X

WHERE PENNYROYAL GREW

WE were a little late to dinner, but Mrs Blackett and Mrs Todd were lenient, and we all took our places after William had paused to wash his hands, like a pious Brahmin, at the well, and put on a neat blue coat which he took from a peg behind the kitchen door. Then he resolutely asked a blessing in words that I could not hear, and we ate the chowder and were thankful. The kitten went round and round the table, quite erect, and, holding on by her fierce young claws, she stopped to mew with pathos at each elbow, or darted off to the open door when a song sparrow forgot himself and lit in the grass too near. William did not talk much, but his sister Todd occupied the time and told all the news there was to tell of Dunnet Landing and its coasts, while the old mother listened with delight. Her hospitality was something exquisite; she had the gift which so many women lack, of being able to make themselves and their houses belong entirely to a guest's pleasure,—that charming surrender for the moment of themselves and whatever belongs to them, so that they make a part of one's own life that can never be forgotten. Tact is after all a kind of mind-reading, and my hostess held the golden gift. Sympathy is of the mind as well as the heart, and Mrs Blackett's world and mine were one from the moment we met. Besides, she had that final, that highest gift of heaven, a perfect self-forgetfulness. Sometimes, as I watched her eager, sweet old face, I wondered why she had been set to shine on this lonely island of the northern coast. It must have been to keep the balance true, and make up to all her scattered and depending neighbors for other things which they may have lacked.

When we had finished clearing away the old blue plates, and the kitten had taken care of her share of the fresh haddock, just as we were putting back the kitchen chairs in their places, Mrs Todd said briskly that she must go up into the pasture now to gather the desired herbs.

'You can stop here an' rest, or you can accompany me,' she announced. 'Mother ought to have her nap, and when we come back she an' William'll sing for you. She admires music,' said Mrs Todd, turning to speak to her mother.

But Mrs Blackett tried to say that she couldn't sing as she used, and perhaps William wouldn't feel like it. She looked tired, the good old soul, or I should have liked to sit in the peaceful little house while she slept; I had had much pleasant experience of pastures already in her daughter's company. But it seemed best to go with Mrs Todd, and off we went.

Mrs Todd carried the gingham bag which she had brought from home, and a small heavy burden in the bottom made it hang straight and slender from her hand. The way was steep, and she soon grew breathless, so that we sat down to rest awhile on a convenient large stone among the bayberry.

'There, I wanted you to see this,—'t is mother's picture,' said Mrs Todd; ''t was taken once when she was up to Portland, soon after she was married. That's me,' she added, opening another worn case, and displaying the full face of the cheerful child she looked like still in spite of being past sixty. 'And here's William an' father together. I take after father, large and heavy, an' William is like mother's folks, short an' thin. He ought to have made something o' himself, bein' a man an' so like mother; but though he's been very steady to work, an' kept up the farm, an' done his fishin' too right along, he never had mother's snap an' power o' seein' things just as they be.* He's got excellent judgment, too,' meditated William's sister, but she could not arrive at any satisfactory decision upon what she evidently thought his failure in life. 'I think it is well to see any one so happy an' makin' the most of life just as it falls to hand,' she said as she began to put the daguerreotypes away again; but I reached out my hand to see her mother's once more, a most flower-like face of a lovely young woman in quaint dress. There was in the eyes a look of anticipation and joy, a far-off look that sought the horizon; one often sees it in seafaring families, inherited by girls and boys alike from men who spend their lives at sea, and are always watching for distant sails or the first loom of the land. At sea there is nothing to be seen close by, and this has its counterpart in a sailor's character, in the large and brave and patient traits that are developed, the hopeful pleasantness that one loves so in a seafarer.

When the family pictures were wrapped again in a big handkerchief, we set forward in a narrow footpath and made our way to a lonely place that faced northward, where there was more pasturage and fewer bushes, and we went down to the edge of short grass above some rocky cliffs where the deep sea broke with a great noise, though the wind was down and the water looked quiet a little way from shore. Among the grass grew such pennyroyal as the rest of the world could not provide. There was a fine fragrance in the air as we gathered it sprig by sprig and stepped along carefully, and Mrs Todd pressed her aromatic nosegay between her hands and offered it to me again and again.

'There's nothin' like it,' she said; 'oh no, there's no such pennyr'yal as this in the State of Maine. It's the right pattern of the plant, and all the rest I ever see is but an imitation.* Don't it do you good?' And I answered with enthusiasm.

'There, dear, I never showed nobody else but mother where to find this place; 't is kind of sainted to me. Nathan, my husband, an' I used to love this place when we was courtin', and'—she hesitated, and then spoke softly—'when he was lost, 't was just off shore tryin' to get in by the short channel out there between Squaw Islands, right in sight o' this headland where we'd set an' made our plans all summer long.'

I had never heard her speak of her husband before, but I felt that we were friends now since she had brought me to this place.

' 'T was but a dream with us,' Mrs Todd said. 'I knew it when he was gone. I knew it'—and she whispered as if she were at confession—'I knew it afore he started to go to sea. My heart was gone out o' my keepin' before I ever saw Nathan; but he loved me well, and he made me real happy, and he died before he ever knew what he'd had to know if we'd lived long together. 'T is very strange about love. No, Nathan never found out, but my heart was troubled when I knew him first. There's more women likes to be loved than there is of those that loves. I spent some happy hours right here. I always liked Nathan, and he never knew. But this pennyr'yal always reminded me, as I'd sit and gather it and hear him talkin''—it always would remind me of—the other one.'

She looked away from me, and presently rose and went on by herself. There was something lonely and solitary about her great determined shape. She might have been Antigone alone on the Theban plain.* It is not often given in a noisy world to come to the places of great grief and silence. An absolute, archaic grief possessed this country-woman; she seemed like a renewal of some historic soul, with her sorrows and the remoteness of a daily life busied with rustic simplicities and the scents of primeval herbs.

I was not incompetent at herb-gathering, and after a while, when I had sat long enough waking myself to new thoughts, and reading a page of remembrance with new pleasure, I gathered some bunches, as I was bound to do, and at last we met again higher up the shore, in the plain every-day world we had left behind when we went down to the pennyroyal plot. As we walked together along the high edge of the field we saw a hundred sails about the bay and farther seaward; it was mid-afternoon or after, and the day was coming to an end.

'Yes, they're all makin' towards the shore,—the small craft an' the lobster smacks an' all,' said my companion. 'We must spend a little time with mother now, just to have our tea, an' then put for home.'

'No matter if we lose the wind at sun-down; I can row in with Johnny,' said I; and Mrs Todd nodded reassuringly and kept to her steady plod, not quickening her gait even when we saw William come round the corner of the house as if to look for us, and wave his hand and disappear.

'Why, William's right on deck; I didn't know 's we should see any more of him!' exclaimed Mrs Todd. 'Now mother'll put the kettle right on; she's got a good fire goin'.' I too could see the blue smoke thicken, and then we both walked a little faster, while Mrs Todd groped in her full bag of herbs to find the daguerreotypes and be ready to put them in their places.

WILLIAM was sitting on the side door step, and the old mother was busy making her tea; she gave into my hand an old flowered-glass tea-caddy.

'William thought you'd like to see this, when he was settin' the table. My father brought it to my mother from the island of Tobago;* an' here's a pair of beautiful mugs that came with it.' She opened the glass door of a little cupboard beside the chimney. 'These I call my best things, dear,' she said. 'You'd laugh to see how we enjoy 'em Sunday nights in winter: we have a real company tea 'stead o' livin' right along just the same, an' I make somethin' good for a s'prise an' put on some o' my preserves, an' we get a-talkin' together an' have real pleasant times.'

Mrs Todd laughed indulgently, and looked to see what I thought of such childishness.

'I wish I could be here some Sunday evening,' said I.

'William an' me'll be talkin' about you an' thinkin' o' this nice day,' said Mrs Blackett affectionately, and she glanced at William, and he looked up bravely and nodded. I began to discover that he and his sister could not speak their deeper feelings before each other.

'Now I want you an' mother to sing,' said Mrs Todd abruptly, with an air of command, and I gave William much sympathy in his evident distress.

'After I've had my cup o' tea, dear,' answered the old hostess cheerfully; and so we sat down and took our cups and made merry while they lasted. It was impossible not to wish to stay on forever at Green Island, and I could not help saying so.

'I'm very happy here, both winter an' summer,' said old Mrs Blackett. 'William an' I never wish for any other home, do we, William? I'm glad you find it pleasant; I wish you'd come an' stay, dear, whenever you feel inclined. But here's Almiry; I always think Providence was kind to plot an' have her husband leave her a good house where she really belonged. She'd been very restless if she'd had to continue here on Green Island. You wanted more scope, didn't you, Almiry, an' to live in a large place where more things grew? Sometimes folks wonders that we don't live together; perhaps we shall some time,' and a shadow of sadness and apprehension flitted across her face. 'The time o' sickness an' failin' has got to come to all. But Almiry's got an herb that's good for everything.' She smiled as she spoke, and looked bright again.

'There's some herb that's good for everybody, except for them that thinks they're sick when they ain't,' announced Mrs Todd, with a truly professional air of finality. 'Come, William, let's have Sweet Home,* an' then mother'll sing Cupid an' the Bee for us.'*

Then followed a most charming surprise. William mastered his timidity and began to sing. His voice was a little faint and frail, like the family daguerreotypes, but it was a tenor voice, and perfectly true and sweet. I have never heard Home, Sweet Home sung as touchingly and seriously as he sang it; he seemed to make it quite new; and when he paused for a moment at the end of the first line and began the next, the old mother joined him and they sang together, she missing only the higher notes, where he seemed to lend his voice to hers for the moment and carry on her very note and air. It was the silent man's real and only means of expression, and one could have listened forever, and have asked for more and more songs of old Scotch and English inheritance and the best that have lived from the ballad music of the war. Mrs Todd kept time visibly, and sometimes audibly, with her ample foot. I saw the tears in her eyes sometimes, when I could see beyond the tears in mine. But at last the songs ended and the time came to say good-by; it was the end of a great pleasure.

Mrs Blackett, the dear old lady, opened the door of her bedroom while Mrs Todd was tying up the herb bag, and William had gone down to get the boat ready and to blow the horn for Johnny Bowden, who had joined a roving boat party who were off the shore lobstering.

I went to the door of the bedroom, and thought how pleasant it looked, with its pink-and-white patchwork quilt and the brown unpainted paneling of its woodwork.

'Come right in, dear,' she said. 'I want you to set down in my old quilted rockin'-chair there by the window; you'll say it's the prettiest view in the house. I set there a good deal to rest me and when I want to read.'

There was a worn red Bible on the light-stand, and Mrs Blackett's heavy silver-bowed glasses; her thimble was on the narrow window-ledge, and folded carefully on the table was a thick striped-cotton shirt that she was making for her son. Those dear old fingers and their loving stitches, that heart which had made the most of everything that needed love! Here was the real home, the heart of the old house on Green Island! I sat in the rocking-chair, and felt that it was a place of peace, the little brown bedroom, and the quiet outlook upon field and sea and sky.

I looked up, and we understood each other without speaking. 'I shall like to think o' your settin' here to-day,' said Mrs Blackett. 'I want you to come again. It has been so pleasant for William.'

The wind served us all the way home, and did not fall or let the sail slacken until we were close to the shore. We had a generous freight of lobsters in the boat, and new potatoes which William had put aboard, and what Mrs Todd proudly called a full 'kag' of prime number one salted mackerel; and when we landed we had to make business arrangements to have these conveyed to her house in a wheelbarrow.

I never shall forget the day at Green Island. The town of Dunnet Landing seemed large and noisy and oppressive as we came ashore. Such is the power of contrast; for the village was so still that I could hear the shy whippoorwills singing that night as I lay awake in my downstairs bedroom, and the scent of Mrs Todd's herb garden under the window blew in again and again with every gentle rising of the sea-breeze.

A STRANGE SAIL

EXCEPT for a few stray guests, islanders or from the inland country, to whom Mrs Todd offered the hospitalities of a single meal, we were quite by ourselves all summer; and when there were signs of invasion, late in July, and a certain Mrs Fosdick appeared like a strange sail on the far horizon, I suffered much from apprehension. I had been living in the quaint little house with as much comfort and unconsciousness as if it were a larger body, or a double shell, in whose simple convolutions Mrs Todd and I had secreted ourselves, until some wandering hermit crab of a visitor marked the little spare room for her own. Perhaps now and then a castaway on a lonely desert island dreads the thought of being rescued. I heard of Mrs Fosdick for the first time with a selfish sense of objection; but after all, I was still vacation-tenant of the schoolhouse, where I could always be alone, and it was impossible not to sympathize with Mrs Todd, who, in spite of some preliminary grumbling, was really delighted with the prospect of entertaining an old friend.

For nearly a month we received occasional news of Mrs Fosdick, who seemed to be making a royal progress from house to house in the inland neighborhood, after the fashion of Queen Elizabeth.* One Sunday after another came and went, disappointing Mrs Todd in the hope of seeing her guest at church and fixing the day for the great visit to begin; but Mrs Fosdick was not ready to commit herself to a date. An assurance of 'some time this week' was not sufficiently definite from a free-footed housekeeper's point of view, and Mrs Todd put aside all herb-gathering plans, and went through the various stages of expectation, provocation, and despair. At last she was ready to believe that Mrs Fosdick must have forgotten her promise and returned to her home, which was vaguely said to be over Thomaston way. But one evening, just as the supper-table was cleared and 'readied up,' and Mrs

Todd had put her large apron over her head and stepped forth for an evening stroll in the garden, the unexpected happened. She heard the sound of wheels, and gave an excited cry to me, as I sat by the window, that Mrs Fosdick was coming right up the street.

'She may not be considerate, but she's dreadful good company,' said Mrs Todd hastily, coming back a few steps from the neighborhood of the gate. 'No, she ain't a mite considerate, but there's a small lobster left over from your tea; yes, it's a real mercy there's a lobster. Susan Fosdick might just as well have passed the compliment o' comin' an hour ago.'

'Perhaps she has had her supper,' I ventured to suggest, sharing the housekeeper's anxiety, and meekly conscious of an inconsiderate appetite for my own supper after a long expedition up the bay. There were so few emergencies of any sort at Dunnet Landing that this one appeared overwhelming.

'No, she's rode 'way over from Nahum Brayton's place. I expect they were busy on the farm, and couldn't spare the horse in proper season. You just sly out an' set the teakittle on again, dear, an' drop in a good han'ful o' chips; the fire's all alive. I'll take her right up to lay off her things, an' she'll be occupied with explanations an' gettin' her bunnit off, so you'll have plenty o' time. She's one I shouldn't like to have find me unprepared.'

Mrs Fosdick was already at the gate, and Mrs Todd now turned with an air of complete surprise and delight to welcome her.

'Why, Susan Fosdick,' I heard her exclaim in a fine unhindered voice, as if she were calling across a field, 'I come near giving of you up! I was afraid you'd gone an' 'portioned out my visit to somebody else. I s'pose you've been to supper?'

'Lor', no, I ain't, Almiry Todd,' said Mrs Fosdick cheerfully, as she turned, laden with bags and bundles, from making her adieux to the boy driver. 'I ain't had a mite o' supper, dear. I've been lottin' all the way on a cup o' that best tea o' yourn,—some o' that Oolong you keep in the little chist. I don't want none o' your useful herbs.'

'I keep that tea for ministers' folks,' gayly responded Mrs

Todd. 'Come right along in, Susan Fosdick. I declare if you ain't the same old sixpence!'

As they came up the walk together, laughing like girls, I fled, full of cares, to the kitchen, to brighten the fire and be sure that the lobster, sole dependence of a late supper, was well out of reach of the cat. There proved to be fine reserves of wild raspberries and bread and butter, so that I regained my composure, and waited impatiently for my own share of this illustrious visit to begin. There was an instant sense of high festivity in the evening air from the moment when our guest had so frankly demanded the Oolong tea.

The great moment arrived. I was formally presented at the stair-foot, and the two friends passed on to the kitchen, where I soon heard a hospitable clink of crockery and the brisk stirring of a tea-cup. I sat in my high-backed rocking-chair by the window in the front room with an unreasonable feeling of being left out, like the child who stood at the gate in Hans Andersen's story.* Mrs Fosdick did not look, at first sight, like a person of great social gifts. She was a serious-looking little bit of an old woman, with a birdlike nod of the head. I had often been told that she was the 'best hand in the world to make a visit,'—as if to visit were the highest of vocations; that everybody wished for her, while few could get her; and I saw that Mrs Todd felt a comfortable sense of distinction in being favored with the company of this eminent person who 'knew just how.' It was certainly true that Mrs Fosdick gave both her hostess and me a warm feeling of enjoyment and expectation, as if she had the power of social suggestion to all neighboring minds.

The two friends did not reappear for at least an hour. I could hear their busy voices, loud and low by turns, as they ranged from public to confidential topics. At last Mrs Todd kindly remembered me and returned, giving my door a ceremonious knock before she stepped in, with the small visitor in her wake. She reached behind her and took Mrs Fosdick's hand as if she were young and bashful, and gave her a gentle pull forward.

'There, I don't know whether you're goin' to take to each other or not; no, nobody can't tell whether you'll suit each other, but I expect you'll get along some way, both having seen the world,' said our affectionate hostess. 'You can inform Mis' Fosdick how we found the folks out to Green Island the other day. She's always been well acquainted with mother. I'll slip out now an' put away the supper things an' set my bread to rise, if you'll both excuse me. You can come out an' keep me company when you get ready, either or both.' And Mrs Todd, large and amiable, disappeared and left us.

Being furnished not only with a subject of conversation, but with a safe refuge in the kitchen in case of incompatibility, Mrs Fosdick and I sat down, prepared to make the best of each other. I soon discovered that she, like many of the elder women of that coast, had spent a part of her life at sea, and was full of a good traveler's curiosity and enlightenment. By the time we thought it discreet to join our hostess we were already sincere friends.

You may speak of a visit's setting in as well as a tide's, and it was impossible, as Mrs Todd whispered to me, not to be pleased at the way this visit was setting in; a new impulse and refreshing of the social currents and seldom visited bays of memory appeared to have begun. Mrs Fosdick had been the mother of a large family of sons and daughters,—sailors and sailors' wives,—and most of them had died before her. I soon grew more or less acquainted with the histories of all their fortunes and misfortunes, and subjects of an intimate nature were no more withheld from my ears than if I had been a shell on the mantelpiece. Mrs Fosdick was not without a touch of dignity and elegance; she was fashionable in her dress, but it was a curiously well-preserved provincial fashion of some years back. In a wider sphere one might have called her a woman of the world, with her unexpected bits of modern knowledge, but Mrs Todd's wisdom was an intimation of truth itself. She might belong to any age, like an idyl of Theocritus;* but while she always understood Mrs Fosdick, that entertaining pilgrim could not always understand Mrs Todd.

That very first evening my friends plunged into a borderless sea of reminiscences and personal news. Mrs Fosdick had been staying with a family who owned the farm where she was born, and she had visited every sunny knoll and shady field corner; but when she said that it might be for the last time, I detected in her tone something expectant of the contradiction which Mrs Todd promptly offered.

'Almiry,' said Mrs Fosdick, with sadness, 'you may say what you like, but I am one of nine brothers and sisters brought up on the old place, and we're all dead but me.'

'Your sister Dailey ain't gone, is she? Why, no, Louisa ain't gone!' exclaimed Mrs Todd, with surprise. 'Why, I never heard of that occurrence!'

'Yes 'm; she passed away last October, in Lynn. She had made her distant home in Vermont State, but she was making a visit to her youngest daughter. Louisa was the only one of my family whose funeral I wasn't able to attend, but 't was a mere accident. All the rest of us were settled right about home. I thought it was very slack of 'em in Lynn not to fetch her to the old place; but when I came to hear about it, I learned that they'd recently put up a very elegant monument, and my sister Dailey was always great for show. She'd just been out to see the monument the week before she was taken down, and admired it so much that they felt sure of her wishes.'

'So she's really gone, and the funeral was up to Lynn!' repeated Mrs Todd, as if to impress the sad fact upon her mind. 'She was some years younger than we be, too. I recollect the first day she ever came to school; 't was that first year mother sent me inshore to stay with aunt Topham's folks and get my schooling. You fetched little Louisa to school one Monday mornin' in a pink dress an' her long curls, and she set between you an' me, and got cryin' after a while, so the teacher sent us home with her at recess.'

'She was scared of seeing so many children about her; there was only her and me and brother John at home then; the older boys were to sea with father, an' the rest of us wa'n't born,' explained Mrs Fosdick. 'That next fall we all went to sea together. Mother was uncertain till the last minute, as one may say. The ship was waiting orders, but the baby that then was, was born just in time, and there was a long spell of extra bad weather, so mother got about again before they had to sail, an' we all went. I remember my clothes were all left ashore in the east chamber in a basket where mother'd took them out o' my chist o' drawers an' left 'em ready to carry aboard. She didn't have nothing aboard, of her own, that she wanted to cut up for me, so when my dress wore out she just put me into a spare suit o' John's, jacket and trousers. I wasn't but eight years old an' he was most seven and large of his age. Quick as we made a port she went right ashore an' fitted me out pretty, but we was bound for the East Indies and didn't put in anywhere for a good while. So I had quite a spell o' freedom. Mother made my new skirt long because I was growing, and I poked about the deck after that, real discouraged, feeling the hem at my heels every minute, and as if youth was past and gone. I liked the trousers best; I used to climb the riggin' with 'em and frighten mother till she said an' vowed she'd never take me to sea again.'

I thought by the polite absent-minded smile on Mrs Todd's face this was no new story.

'Little Louisa was a beautiful child; yes, I always thought Louisa was very pretty,' Mrs Todd said. 'She was a dear little girl in those days. She favored your mother; the rest of you took after your father's folks.'

'We did certain,' agreed Mrs Fosdick, rocking steadily. 'There, it does seem so pleasant to talk with an old acquaintance that knows what you know. I see so many of these new folks nowadays, that seem to have neither past nor future. Conversation's got to have some root in the past, or else you've got to explain every remark you make, an' it wears a person out.'

Mrs Todd gave a funny little laugh. 'Yes 'm, old friends is always best, 'less you can catch a new one that's fit to make an old one out of,' she said, and we gave an affectionate glance at each other which Mrs Fosdick could not have understood, being the latest comer to the house.

XIII

POOR JOANNA

ONE evening my ears caught a mysterious allusion which Mrs Todd made to Shell-heap Island. It was a chilly night of cold northeasterly rain, and I made a fire for the first time in the Franklin stove in my room, and begged my two housemates to come in and keep me company. The weather had convinced Mrs Todd that it was time to make a supply of cough-drops, and she had been bringing forth herbs from dark and dry hiding-places, until now the pungent dust and odor of them had resolved themselves into one mighty flavor of spearmint that came from a simmering caldron of syrup in the kitchen. She called it done, and well done, and had ostentatiously left it to cool, and taken her knitting-work because Mrs Fosdick was busy with hers. They sat in the two rocking-chairs, the small woman and the large one, but now and then I could see that Mrs Todd's thoughts remained with the cough-drops. The time of gathering herbs was nearly over, but the time of syrups and cordials had begun.

The heat of the open fire made us a little drowsy, but something in the way Mrs Todd spoke of Shell-heap Island waked my interest. I waited to see if she would say any more, and then took a roundabout way back to the subject by saying what was first in my mind: that I wished the Green Island family were there to spend the evening with us,—Mrs Todd's mother and her brother William.

Mrs Todd smiled, and drummed on the arm of the rocking-chair. 'Might scare William to death,' she warned me; and Mrs Fosdick mentioned her intention of going out to Green Island to stay two or three days, if this wind didn't make too much sea.

'Where is Shell-heap Island?' I ventured to ask, seizing the opportunity.

'Bears nor'east somewheres about three miles from Green Island; right off-shore, I should call it about eight miles out,' said Mrs Todd. 'You never was there, dear; 't is off the thoroughfares, and a very bad place to land at best.'

'I should think 't was,' agreed Mrs Fosdick, smoothing down her black silk apron. ' 'T is a place worth visitin' when you once get there. Some o' the old folks was kind o' fearful about it. 'T was 'counted a great place in old Indian times; you can pick up their stone tools 'most any time if you hunt about. There's a beautiful spring o' water, too. Yes, I remember when they used to tell queer stories about Shell-heap Island. Some said 't was a great bangeing-place* for the Indians, and an old chief resided there once that ruled the winds; and others said they'd always heard that once the Indians come down from up country an' left a captive there without any bo't, an' 't was too far to swim across to Black Island, so called, an' he lived there till he perished.'

'I've heard say he walked the island after that, and sharp-sighted folks could see him an' lose him like one o' them citizens Cap'n Littlepage was acquainted with up to the north pole,' announced Mrs Todd grimly. 'Anyway, there was Indians,—you can see their shell-heap that named the island; and I've heard myself that 't was one o' their cannibal places, but I never could believe it. There never was no cannibals on the coast o' Maine. All the Indians o' these regions are tame-looking folks.'

'Sakes alive, yes!' exclaimed Mrs Fosdick. 'Ought to see them painted savages I've seen when I was young out in the South Sea Islands! That was the time for folks to travel, 'way back in the old whalin' days!'

'Whalin' must have been dull for a lady, hardly ever makin' a lively port, and not takin' in any mixed cargoes,' said Mrs Todd. 'I never desired to go a whalin' v'y'ge myself.'

'I used to return feelin' very slack an' behind the times, 't is true,' explained Mrs Fosdick, 'but 't was excitin', an' we always done extra well, and felt rich when we did get ashore. I liked the variety. There, how times have changed; how few seafarin' families there are left! What a lot o' queer folks there used to be about here, anyway, when we was young, Almiry. Everybody's just like everybody else, now; nobody to laugh about, and nobody to cry about.'

It seemed to me that there were peculiarities of character in the region of Dunnet Landing yet, but I did not like to interrupt.

'Yes,' said Mrs Todd after a moment of meditation, 'there was certain a good many curiosities of human natur' in this neighborhood years ago. There was more energy then, and in some the energy took a singular turn. In these days the young folks is all copy-cats, 'fraid to death they won't be all just alike; as for the old folks, they pray for the advantage o' bein' a little different.'

'I ain't heard of a copy-cat this great many years,' said Mrs Fosdick, laughing; ''t was a favorite term o' my grand-mother's. No, I wa'n't thinking o' those things, but of them strange straying creatur's that used to rove the country. You don't see them now, or the ones that used to hive away in their own houses with some strange notion or other.'

I thought again of Captain Littlepage, but my companions were not reminded of his name; and there was brother William at Green Island, whom we all three knew.

'I was talking o' poor Joanna the other day. I hadn't thought of her for a great while,' said Mrs Fosdick abruptly. 'Mis' Brayton an' I recalled her as we sat together sewing. She was one o' your peculiar persons, wa'n't she? Speaking of such persons,' she turned to explain to me, 'there was a sort of a nun or hermit person lived out there for years all alone on Shell-heap Island. Miss Joanna Todd, her name was,—a cousin o' Almiry's late husband.'

I expressed my interest, but as I glanced at Mrs Todd I saw that she was confused by sudden affectionate feeling and unmistakable desire for reticence.

'I never want to hear Joanna laughed about,' she said anxiously.

'Nor I,' answered Mrs Fosdick reassuringly. 'She was crossed in love,—that was all the matter to begin with; but as I look back, I can see that Joanna was one doomed from the first to fall into a melancholy. She retired from the world for good an' all, though she was a well-off woman. All she wanted was to get away from folks; she thought she wasn't fit to live with anybody, and wanted to be free. Shell-heap Island come to her from her father, and first thing folks knew she'd gone off out there to live, and left word she didn't want no company. 'T was a bad place to get to, unless the wind an' tide were just right; 't was hard work to make a landing.'

'What time of year was this?' I asked.

'Very late in the summer,' said Mrs Fosdick. 'No, I never could laugh at Joanna, as some did. She set everything by the young man, an' they were going to marry in about a month, when he got bewitched with a girl 'way up the bay, and married her, and went off to Massachusetts. He wasn't well thought of,—there were those who thought Joanna's money was what had tempted him; but she'd given him her whole heart, an' she wa'n't so young as she had been. All her hopes were built on marryin', an' havin' a real home and somebody to look to; she acted just like a bird when its nest is spoilt. The day after she heard the news she was in dreadful woe, but the next she came to herself very quiet, and took the horse and wagon, and drove fourteen miles to the lawyer's, and signed a paper givin' her half of the farm to her brother. They never had got along very well together, but he didn't want to sign it, till she acted so distressed that he gave in. Edward Todd's wife was a good woman, who felt very bad indeed, and used every argument with Joanna; but Joanna took a poor old boat that had been her father's and lo'ded in a few things, and off she put all alone, with a good land breeze, right out to sea. Edward Todd ran down to the beach, an' stood there cryin' like a boy to see her go, but she was out o' hearin'. She never stepped foot on the mainland again long as she lived.'

'How large an island is it? How did she manage in winter?' I asked.

'Perhaps thirty acres, rocks and all,' answered Mrs Todd, taking up the story gravely. 'There can't be much of it that the salt spray don't fly over in storms. No, 't is a dreadful small place to make a world of; it has a different look from any of the other islands, but there's a sheltered cove on the south side, with mud-flats across one end of it at low water where there's excellent clams, and the big shell-heap keeps some o' the wind off a little house her father took the trouble to build when he was a young man. They said there was an old house built o' logs there before that, with a kind of natural cellar in the rock under it. He used to stay out there days to a time, and anchor a little sloop he had, and dig clams to fill it, and sail up to Portland. They said the dealers always gave him an extra price, the clams were so noted. Joanna used to go out and stay with him. They were always great companions, so she knew just what 't was out there. There was a few sheep that belonged to her brother an' her, but she bargained for him to come and get them on the edge o' cold weather. Yes, she desired him to come for the sheep; an' his wife thought perhaps Joanna'd return, but he said no, an' lo'ded the bo't with warm things an' what he thought she'd need through the winter. He come home with the sheep an' left the other things by the house, but she never so much as looked out o' the window. She done it for a penance. She must have wanted to see Edward by that time.'

Mrs Fosdick was fidgeting with eagerness to speak.

'Some thought the first cold snap would set her ashore, but she always remained,' concluded Mrs Todd soberly.

'Talk about the men not having any curiosity!' exclaimed Mrs Fosdick scornfully. 'Why, the waters round Shell-heap Island were white with sails all that fall. 'T was never called no great of a fishin'-ground before. Many of 'em made excuse to go ashore to get water at the spring; but at last she spoke to a bo't-load, very dignified and calm, and said that she'd like it better if they'd make a practice of getting water to Black Island or somewheres else and leave her alone, except in case of accident or trouble. But there was one man who had always set everything by her from a boy. He'd have married her if the other hadn't come about an' spoilt his chance, and he used to get close to the island, before light, on his way out fishin', and throw a little bundle 'way up the green slope front o' the house. His sister told me she happened to see, the first time, what a pretty choice he made o' useful things that a woman would feel lost without. He stood off fishin', and could see them in the grass all day, though sometimes she'd come out and walk right by them. There was other bo'ts near, out after mackerel. But early next morning his present was gone. He didn't presume too much, but once he took her a nice firkin o' things he got up to Portland, and when spring come he landed her a hen and chickens in a nice little coop. There was a good many old friends had Joanna on their minds.'

'Yes,' said Mrs Todd, losing her sad reserve in the growing sympathy of these reminiscences. 'How everybody used to notice whether there was smoke out of the chimney! The Black Island folks could see her with their spy-glass, and if they'd ever missed getting some sign o' life they'd have sent notice to her folks. But after the first year or two Joanna was more and more forgotten as an every-day charge. Folks lived very simple in those days, you know,' she continued, as Mrs Fosdick's knitting was taking much thought at the moment. 'I expect there was always plenty of driftwood thrown up, and a poor failin' patch of spruces covered all the north side of the island, so she always had something to burn. She was very fond of workin' in the garden ashore, and that first summer she began to till the little field out there, and raised a nice parcel o' potatoes. She could fish, o' course, and there was all her clams an' lobsters. You can always live well in any wild place by the sea when you'd starve to death up country, except 't was berry time. Joanna had berries out there, black-berries at least, and there was a few herbs in case she needed them. Mullein in great quantities and a plant o' wormwood I remember seeing once when I stayed there, long before she fled out to Shell-heap. Yes, I recall the wormwood, which is always a planted herb, so there must have been folks there before the Todds' day. A growin' bush makes the best grave-stone; I expect that wormwood always stood for somebody's solemn monument. Catnip, too, is a very endurin' herb about an old place.'

'But what I want to know is what she did for other things,' interrupted Mrs Fosdick. 'Almiry, what did she do for clothin' when she needed to replenish, or risin' for her bread, or the piece-bag that no woman can live long without?'

'Or company,' suggested Mrs Todd. 'Joanna was one that loved her friends. There must have been a terrible sight o' long winter evenin's that first year.'

'There was her hens,' suggested Mrs Fosdick, after review-ing the melancholy situation. 'She never wanted the sheep after that first season. There wa'n't no proper pasture for sheep after the June grass was past, and she ascertained the fact and couldn't bear to see them suffer; but the chickens done well. I remember sailin' by one spring afternoon, an' seein' the coops out front o' the house in the sun. How long was it before you went out with the minister? You were the first ones that ever really got ashore to see Joanna.'

I had been reflecting upon a state of society which admitted such personal freedom and a voluntary hermitage. There was something mediæval in the behavior of poor Joanna Todd under a disappointment of the heart. The two women had drawn closer together, and were talking on, quite unconscious of a listener.

'Poor Joanna!' said Mrs Todd again, and sadly shook her head as if there were things one could not speak about.

'I called her a great fool,' declared Mrs Fosdick, with spirit, 'but I pitied her then, and I pity her far more now. Some other minister would have been a great help to her,—one that preached self-forgetfulness and doin' for others to cure our own ills; but Parson Dimmick was a vague person, well meanin', but very numb in his feelin's. I don't suppose at that troubled time Joanna could think of any way to mend her troubles except to run off and hide.'

'Mother used to say she didn't see how Joanna lived without having nobody to do for, getting her own meals and tending her own poor self day in an' day out,' said Mrs Todd sorrowfully.

'There was the hens,' repeated Mrs Fosdick kindly. 'I expect she soon came to makin' folks o' them. No, I never went to work to blame Joanna, as some did. She was full o' feeling, and her troubles hurt her more than she could bear. I see it all now as I couldn't when I was young.'

'I suppose in old times they had their shut-up convents for just such folks,' said Mrs Todd, as if she and her friend had disagreed about Joanna once, and were now in happy harmony. She seemed to speak with new openness and freedom. 'Oh yes, I was only too pleased when the Reverend Mr Dimmick invited me to go out with him. He hadn't been very long in the place when Joanna left home and friends. 'T was one day that next summer after she went, and I had been married early in the spring. He felt that he ought to go out and visit her. She was a member of the church, and might wish to have him consider her spiritual state. I wa'n't so sure o' that, but I always liked Joanna, and I'd come to be her cousin by marriage. Nathan an' I had conversed about goin' out to pay her a visit, but he got his chance to sail sooner 'n he expected. He always thought everything of her, and last time he come home, knowing nothing of her change, he brought her a beautiful coral pin from a port he'd touched at somewheres up the Mediterranean. So I wrapped the little box in a nice piece of paper and put it in my pocket, and picked her a bunch of fresh lemon balm, and off we started.'

Mrs Fosdick laughed. 'I remember hearin' about your trials on the v'y'ge,' she said.

'Why, yes,' continued Mrs Todd in her company manner. 'I picked her the balm, an' we started. Why, yes, Susan, the minister liked to have cost me my life that day. He would fasten the sheet, though I advised against it. He said the rope was rough an' cut his hand. There was a fresh breeze, an' he went on talking rather high flown, an' I felt some interested. All of a sudden there come up a gust, and he give a screech and stood right up and called for help, 'way out there to sea. I knocked him right over into the bottom o' the bo't, getting by to catch hold of the sheet an' untie it. He wasn't but a little man; I helped him right up after the squall passed, and made a handsome apology to him, but he did act kind o' offended.'

'I do think they ought not to settle them landlocked folks in parishes where they're liable to be on the water,' insisted Mrs Fosdick. 'Think of the families in our parish that was scattered all about the bay, and what a sight o' sails you used to see, in Mr Dimmick's day, standing across to the mainland on a pleasant Sunday morning, filled with church-going folks, all sure to want him some time or other! You couldn't find no doctor that would stand up in the boat and screech if a flaw struck her.'

'Old Dr Bennett had a beautiful sailboat, didn't he?' responded Mrs Todd. 'And how well he used to brave the weather! Mother always said that in time o' trouble that tall white sail used to look like an angel's wing comin' over the sea to them that was in pain. Well, there's a difference in gifts. Mr Dimmick was not without light.'

' 'T was light o' the moon, then,' snapped Mrs Fosdick; 'he was pompous enough, but I never could remember a single word he said. There, go on, Mis' Todd; I forget a great deal about that day you went to see poor Joanna.'

'I felt she saw us coming, and knew us a great way off; yes, I seemed to feel it within me,' said our friend, laying down her knitting. 'I kept my seat, and took the bo't inshore without saying a word; there was a short channel that I was sure Mr Dimmick wasn't acquainted with, and the tide was very low. She never came out to warn us off nor anything, and I thought, as I hauled the bo't up on a wave and let the Reverend Mr Dimmick step out, that it was somethin' gained to be safe ashore. There was a little smoke out o' the chimney o' Joanna's house, and it did look sort of homelike and pleasant with wild mornin'-glory vines trained up; an' there was a plot o' flowers under the front window, portulacas and things. I believe she'd made a garden once, when she was stopping there with her father, and some things must have seeded in. It looked as if she might have gone over to the other side of the island. 'T was neat and pretty all about the house, and a lovely day in July. We walked up from the beach together very sedate, and I felt for poor Nathan's little pin to see if 't was safe in my dress pocket. All of a sudden Joanna come right to the fore door and stood there, not sayin' a word.'

XIV

THE HERMITAGE

MY companions and I had been so intent upon the subject of the conversation that we had not heard any one open the gate, but at this moment, above the noise of the rain, we heard a loud knocking. We were all startled as we sat by the fire, and Mrs Todd rose hastily and went to answer the call, leaving her rocking-chair in violent motion. Mrs Fosdick and I heard an anxious voice at the door speaking of a sick child, and Mrs Todd's kind, motherly voice inviting the messenger in: then we waited in silence. There was a sound of heavy dropping of rain from the eaves, and the distant roar and undertone of the sea. My thoughts flew back to the lonely woman on her outer island; what separation from humankind she must have felt, what terror and sadness, even in a summer storm like this!

'You send right after the doctor if she ain't better in half an hour,' said Mrs Todd to her worried customer as they parted; and I felt a warm sense of comfort in the evident resources of even so small a neighborhood, but for the poor hermit Joanna there was no neighbor on a winter night.

'How did she look?' demanded Mrs Fosdick, without preface, as our large hostess returned to the little room with a mist about her from standing long in the wet doorway, and the sudden draught of her coming beat out the smoke and flame from the Franklin stove. 'How did poor Joanna look?'

'She was the same as ever, except I thought she looked smaller,' answered Mrs Todd after thinking a moment; perhaps it was only a last considering thought about her patient. 'Yes, she was just the same, and looked very nice, Joanna did. I had been married since she left home, an' she treated me like her own folks. I expected she'd look strange, with her hair turned gray in a night or somethin', but she wore a pretty gingham dress I'd often seen her wear before she went away; she must have kept it nice for best in the afternoons. She always had beautiful, quiet manners. I remember she waited till we were close to her, and then kissed me real affectionate, and inquired for Nathan before she shook hands with the minister, and then she invited us both in. 'T was the same little house her father had built him when he was a bachelor, with one livin'-room, and a little mite of a bedroom out of it where she slept, but 't was neat as a ship's cabin. There was some old chairs, an' a seat made of a long box that might have held boat tackle an' things to lock up in his fishin' days, and a good enough stove so anybody could cook and keep warm in cold weather. I went over once from home and stayed 'most a week with Joanna when we was girls, and those young happy days rose up before me. Her father was busy all day fishin' or clammin'; he was one o' the pleasantest men in the world, but Joanna's mother had the grim streak, and never knew what 't was to be happy. The first minute my eyes fell upon Joanna's face that day I saw how she had grown to look like Mis' Todd. 'T was the mother right over again.'

'Oh dear me!' said Mrs Fosdick.

'Joanna had done one thing very pretty. There was a little piece o' swamp on the island where good rushes grew plenty, and she'd gathered 'em, and braided some beautiful mats for the floor and a thick cushion for the long bunk. She'd showed a good deal of invention; you see there was a nice chance to pick up pieces o' wood and boards that drove ashore, and she'd made good use o' what she found. There wasn't no clock, but she had a few dishes on a shelf, and flowers set about in shells fixed to the walls, so it did look sort of home-like, though so lonely and poor. I couldn't keep the tears out o' my eyes, I felt so sad. I said to myself, I must get mother to come over an' see Joanna; the love in mother's heart would warm her, an' she might be able to advise.'

'Oh no, Joanna was dreadful stern,' said Mrs Fosdick.

'We were all settin' down very proper, but Joanna would keep stealin' glances at me as if she was glad I come. She had but little to say; she was real polite an' gentle, and yet forbiddin'. The minister found it hard,' confessed Mrs Todd; 'he got embarrassed, an' when he put on his authority and asked

her if she felt to enjoy religion in her present situation, an'
she replied that she must be excused from answerin', I
thought I should fly. She might have made it easier for him;
after all, he was the minister and had taken some trouble to
come out, though 't was kind of cold an' unfeelin' the way he
inquired. I thought he might have seen the little old Bible a-
layin' on the shelf close by him, an' I wished he knew enough
to just lay his hand on it an' read somethin' kind an' fatherly
'stead of accusin' her, an' then given poor Joanna his blessin'
with the hope she might be led to comfort. He did offer
prayer, but 't was all about hearin' the voice o' God out o' the
whirlwind;* and I thought while he was goin' on that anybody
that had spent the long cold winter all alone out on Shell-
heap Island knew a good deal more about those things than
he did. I got so provoked I opened my eyes and stared right at
him.

'She didn't take no notice, she kep' a nice respectful manner towards him, and when there come a pause she asked if
he had any interest about the old Indian remains, and took
down some queer stone gouges and hammers off of one of
her shelves and showed them to him same 's if he was a boy.
He remarked that he'd like to walk over an' see the shell-
heap; so she went right to the door and pointed him the way.
I see then that she'd made her some kind o' sandal-shoes out
o' the fine rushes to wear on her feet; she stepped light an'
nice in 'em as shoes.'

Mrs Fosdick leaned back in her rocking-chair and gave a
heavy sigh.

'I didn't move at first, but I'd held out just as long as I
could,' said Mrs Todd, whose voice trembled a little. 'When
Joanna returned from the door, an' I could see that man's
stupid back departin' among the wild rose bushes, I just ran
to her an' caught her in my arms. I wasn't so big as I be now,
and she was older than me, but I hugged her tight, just as if
she was a child. "Oh, Joanna dear," I says, "won't you come
ashore an' live 'long o' me at the Landin', or go over to Green
Island to mother's when winter comes? Nobody shall trouble
you, an' mother finds it hard bein' alone. I can't bear to leave

you here"—and I burst right out crying. I'd had my own trials, young as I was, an' she knew it. Oh, I did entreat her; yes, I entreated Joanna.'

'What did she say then?' asked Mrs Fosdick, much moved.

'She looked the same way, sad an' remote through it all,' said Mrs Todd mournfully. 'She took hold of my hand, and we sat down close together; 't was as if she turned round an' made a child of me. "I haven't got no right to live with folks no more," she said. "You must never ask me again, Almiry: I've done the only thing I could do, and I've made my choice. I feel a great comfort in your kindness, but I don't deserve it. I have committed the unpardonable sin;* you don't under-stand," says she humbly. "I was in great wrath and trouble, and my thoughts was so wicked towards God that I can't expect ever to be forgiven. I have come to know what it is to have patience, but I have lost my hope. You must tell those that ask how 't is with me," she said, "an' tell them I want to be alone." I couldn't speak; no, there wa'n't anything I could say, she seemed so above everything common. I was a good deal younger then than I be now, and I got Nathan's little coral pin out o' my pocket and put it into her hand; and when she saw it and I told her where it come from, her face did really light up for a minute, sort of bright an' pleasant. "Nathan an' I was always good friends; I'm glad he don't think hard of me," says she. "I want you to have it, Almiry, an' wear it for love o' both o' us," and she handed it back to me. "You give my love to Nathan,—he's a dear good man," she said; "an' tell your mother, if I should be sick she mustn't wish I could get well, but I want her to be the one to come." Then she seemed to have said all she wanted to, as if she was done with the world, and we sat there a few minutes longer together. It was real sweet and quiet except for a good many birds and the sea rollin' up on the beach; but at last she rose, an' I did too, and she kissed me and held my hand in hers a minute, as if to say good-by; then she turned and went right away out o' the door and disappeared.

'The minister come back pretty soon, and I told him I was all ready, and we started down to the bo't. He had picked up some round stones and things and was carrying them in his pocket-handkerchief; an' he sat down amidships without making any question, and let me take the rudder an' work the bo't, an' made no remarks for some time, until we sort of eased it off speaking of the weather, an' subjects that arose as we skirted Black Island, where two or three families lived belongin' to the parish. He preached next Sabbath as usual, somethin' high soundin' about the creation, and I couldn't help thinkin' he might never get no further; he seemed to know no remedies, but he had a great use of words.'

Mrs Fosdick sighed again. 'Hearin' you tell about Joanna brings the time right back as if 't was yesterday,' she said. 'Yes, she was one o' them poor things that talked about the great sin; we don't seem to hear nothing about the unpardonable sin now, but you may say 't was not uncommon then.'

'I expect that if it had been in these days, such a person would be plagued to death with idle folks,' continued Mrs Todd, after a long pause. 'As it was, nobody trespassed on her; all the folks about the bay respected her an' her feelings; but as time wore on, after you left here, one after another ventured to make occasion to put somethin' ashore for her if they went that way. I know mother used to go to see her sometimes, and send William over now and then with something fresh an' nice from the farm. There is a point on the sheltered side where you can lay a boat close to shore an' land anything safe on the turf out o' reach o' the water. There were one or two others, old folks, that she would see, and now an' then she'd hail a passin' boat an' ask for somethin'; and mother got her to promise that she would make some sign to the Black Island folks if she wanted help. I never saw her myself to speak to after that day.'

'I expect nowadays, if such a thing happened, she'd have gone out West to her uncle's folks or up to Massachusetts and had a change, an' come home good as new. The world's bigger an' freer than it used to be,' urged Mrs Fosdick.

'No,' said her friend. ''T is like bad eyesight, the mind of such a person: if your eyes don't see right there may be a remedy, but there's no kind of glasses to remedy the mind. No, Joanna was Joanna, and there she lays on her island where she lived and did her poor penance. She told mother the day she was dyin' that she always used to want to be fetched inshore when it come to the last; but she'd thought it over, and desired to be laid on the island, if 't was thought right. So the funeral was out there, a Saturday afternoon in September. 'T was a pretty day, and there wa'n't hardly a boat on the coast within twenty miles that didn't head for Shell-heap cram-full o' folks, an' all real respectful, same 's if she'd always stayed ashore and held her friends. Some went out o' mere curiosity, I don't doubt,—there's always such to every funeral; but most had real feelin', and went purpose to show it. She'd got most o' the wild sparrows as tame as could be, livin' out there so long among 'em, and one flew right in and lit on the coffin an' begun to sing while Mr Dimmick was speakin'. He was put out by it, an' acted as if he didn't know whether to stop or go on. I may have been prejudiced, but I wa'n't the only one thought the poor little bird done the best of the two.'

'What became o' the man that treated her so, did you ever hear?' asked Mrs Fosdick. 'I know he lived up to Mass-achusetts for a while. Somebody who came from the same place told me that he was in trade there an' doin' very well, but that was years ago.'

'I never heard anything more than that; he went to the war in one o' the early rigiments. No, I never heard any more of him,' answered Mrs Todd. 'Joanna was another sort of per-son, and perhaps he showed good judgment in marryin' somebody else, if only he'd behaved straight-forward and manly. He was a shifty-eyed, coaxin' sort of man, that got what he wanted out o' folks, an' only gave when he wanted to buy, made friends easy and lost 'em without knowin' the differ-ence. She'd had a piece o' work tryin' to make him walk accordin' to her right ideas, but she'd have had too much variety ever to fall into a melancholy. Some is meant to be the Joannas in this world, an' 't was her poor lot.'

XV

ON SHELL-HEAP ISLAND

SOME time after Mrs Fosdick's visit was over and we had returned to our former quietness, I was out sailing alone with Captain Bowden in his large boat. We were taking the crooked northeasterly channel seaward, and were well out from shore while it was still early in the afternoon. I found myself presently among some unfamiliar islands, and suddenly remembered the story of poor Joanna. There is something in the fact of a hermitage that cannot fail to touch the imagination; the recluses are a sad kindred, but they are never commonplace. Mrs Todd had truly said that Joanna was like one of the saints in the desert;* the loneliness of sorrow will forever keep alive their sad succession.

'Where is Shell-heap Island?' I asked eagerly.

'You see Shell-heap now, layin' 'way out beyond Black Island there,' answered the captain, pointing with outstretched arm as he stood, and holding the rudder with his knee.

'I should like very much to go there,' said I, and the captain, without comment, changed his course a little more to the eastward and let the reef out of his mainsail.

'I don't know 's we can make an easy landin' for ye,' he remarked doubtfully. 'May get your feet wet; bad place to land. Trouble is I ought to have brought a tag-boat; but they clutch on to the water so, an' I do love to sail free. This gre't boat gets easy bothered with anything trailin'. 'T ain't breakin' much on the meetin'-house ledges; guess I can fetch in to Shell-heap.'

'How long is it since Miss Joanna Todd died?' I asked, partly by way of explanation.

'Twenty-two years come September,' answered the captain, after reflection. 'She died the same year my oldest boy was born, an' the town house was burnt over to the Port. I didn't know but you merely wanted to hunt for some o' them Indian relics. Long 's you want to see where Joanna lived—No, 't ain't breakin' over the ledges; we'll manage to fetch across the shoals somehow, 't is such a distance to go 'way round, and tide's a-risin',' he ended hopefully, and we sailed steadily on, the captain speechless with intent watching of a difficult course, until the small island with its low whitish promontory lay in full view before us under the bright afternoon sun.

The month was August, and I had seen the color of the islands change from the fresh green of June to a sunburnt brown that made them look like stone, except where the dark green of the spruces and fir balsam kept the tint that even winter storms might deepen, but not fade. The few wind-bent trees on Shell-heap Island were mostly dead and gray, but there were some low-growing bushes, and a stripe of light green ran along just above the shore, which I knew to be wild morning-glories. As we came close I could see the high stone walls of a small square field, though there were no sheep left to assail it; and below, there was a little harbor-like cove where Captain Bowden was boldly running the great boat in to seek a landing-place. There was a crooked channel of deep water which led close up against the shore.

'There, you hold fast for'ard there, an' wait for her to lift on the wave. You'll make a good landin' if you're smart; right on the port-hand side!' the captain called excitedly; and I, standing ready with high ambition, seized my chance and leaped over to the grassy bank.

'I'm beat if I ain't aground after all!' mourned the captain despondently.

But I could reach the bowsprit, and he pushed with the boat-hook, while the wind veered round a little as if on purpose and helped with the sail; so presently the boat was free and began to drift out from shore.

'Used to call this p'int Joanna's wharf privilege, but 't has worn away in the weather since her time. I thought one or two bumps wouldn't hurt us none,—paint's got to be renewed, anyway,—but I never thought she'd tetch. I figured on shyin' by,' the captain apologized. 'She's too gre't a boat to handle well in here; but I used to sort of shy by in Joanna's day, an' cast a little somethin' ashore—some apples or a couple o' pears if I had 'em—on the grass, where she'd be sure to see.'

I stood watching while Captain Bowden cleverly found his

way back to deeper water. 'You needn't make no haste,' he called to me; 'I'll keep within call. Joanna lays right up there in the far corner o' the field. There used to be a path led to the place. I always knew her well. I was out here to the funeral.'

I found the path; it was touching to discover that this lonely spot was not without its pilgrims. Later generations will know less and less of Joanna herself, but there are paths trodden to the shrines of solitude the world over,—the world cannot forget them, try as it may; the feet of the young find them out because of curiosity and dim foreboding, while the old bring hearts full of remembrance. This plain anchorite had been one of those whom sorrow made too lonely to brave the sight of men, too timid to front the simple world she knew, yet valiant enough to live alone with her poor insistent human nature and the calms and passions of the sea and sky.

The birds were flying all about the field; they fluttered up out of the grass at my feet as I walked along, so tame that I liked to think they kept some happy tradition from summer to summer of the safety of nests and good fellowship of mankind. Poor Joanna's house was gone except the stones of its foundations, and there was little trace of her flower garden except a single faded sprig of much-enduring French pinks, which a great bee and a yellow butterfly were befriending together. I drank at the spring, and thought that now and then some one would follow me from the busy, hard-worked, and simple-thoughted countryside of the mainland, which lay dim and dreamlike in the August haze, as Joanna must have watched it many a day. There was the world, and here was she with eternity well begun. In the life of each of us, I said to myself, there is a place remote and islanded,* and given to endless regret or secret happiness; we are each the uncompanioned hermit and recluse of an hour or a day; we understand our fellows of the cell to whatever age of history they may belong.

But as I stood alone on the island, in the sea-breeze, suddenly there came a sound of distant voices; gay voices and laughter from a pleasure-boat that was going seaward full of boys and girls. I knew, as if she had told me, that poor Joanna must have heard the like on many and many a summer afternoon, and must have welcomed the good cheer in spite of hopelessness and winter weather, and all the sorrow and disappointment in the world.

XVI

THE GREAT EXPEDITION

MRS TODD never by any chance gave warning over night of her great projects and adventures by sea and land. She first came to an understanding with the primal forces of nature, and never trusted to any preliminary promise of good weather, but examined the day for herself in its infancy. Then, if the stars were propitious, and the wind blew from a quarter of good inheritance whence no surprises of sea-turns or southwest sultriness might be feared, long before I was fairly awake I used to hear a rustle and knocking like a great mouse in the walls, and an impatient tread on the steep garret stairs that led to Mrs Todd's chief place of storage. She went and came as if she had already started on her expedition with utmost haste and kept returning for something that was forgotten. When I appeared in quest of my breakfast, she would be absent-minded and sparing of speech, as if I had displeased her, and she was now, by main force of principle, holding herself back from altercation and strife of tongues.

These signs of a change became familiar to me in the course of time, and Mrs Todd hardly noticed some plain proofs of divination one August morning when I said, without preface, that I had just seen the Beggs' best chaise go by, and that we should have to take the grocery. Mrs Todd was alert in a moment.

'There! I might have known!' she exclaimed. 'It's the 15th of August, when he goes and gets his money. He heired an annuity from an uncle o' his on his mother's side. I understood the uncle said none o' Sam Begg's wife's folks should make free with it, so after Sam's gone it'll all be past an' spent, like last summer. That's what Sam prospers on now, if you can call it prosperin'. Yes, I might have known. 'T is the 15th o' August with him, an' he gener'ly stops to dinner with a cousin's widow on the way home. Feb'uary an' August is the times. Takes him 'bout all day to go an' come.'

I heard this explanation with interest. The tone of Mrs Todd's voice was complaining at the last.

'I like the grocery just as well as the chaise,' I hastened to say, referring to a long-bodied high wagon with a canopy-top, like an attenuated four-posted bedstead on wheels, in which we sometimes journeyed. 'We can put things in behind— roots and flowers and raspberries, or anything you are going after—much better than if we had the chaise.'

Mrs Todd looked stony and unwilling. 'I counted upon the chaise,' she said, turning her back to me, and roughly pushing back all the quiet tumblers on the cupboard shelf as if they had been impertinent. 'Yes, I desired the chaise for once. I ain't goin' berryin' nor to fetch home no more wilted vegetation this year. Season's about past, except for a poor few o' late things,' she added in a milder tone. 'I'm goin' up country. No, I ain't intendin' to go berryin'. I've been plottin' for it the past fortnight and hopin' for a good day.'

'Would you like to have me go too?' I asked frankly, but not without a humble fear that I might have mistaken the purpose of this latest plan.

'Oh certain, dear!' answered my friend affectionately. 'Oh no, I never thought o' any one else for comp'ny, if it's convenient for you, long 's poor mother ain't come. I ain't nothin' like so handy with a conveyance as I be with a good bo't. Comes o' my early bringing-up. I expect we've got to make that great high wagon do. The tires want settin'* and 't is all loose-jointed, so I can hear it shackle the other side o' the ridge. We'll put the basket in front. I ain't goin' to have it bouncin' an' twirlin' all the way. Why, I've been makin' some nice hearts and rounds* to carry.'

These were signs of high festivity, and my interest deepened moment by moment.

'I'll go down to the Beggs' and get the horse just as soon as I finish my breakfast,' said I. 'Then we can start whenever you are ready.'

Mrs Todd looked cloudy again. 'I don't know but you look nice enough to go just as you be,' she suggested doubtfully. 'No, you wouldn't want to wear that pretty blue dress o' yourn 'way up country. 'T ain't dusty now, but it may be comin' home. No, I expect you'd rather not wear that and the other hat.'

'Oh yes. I shouldn't think of wearing these clothes,' said I, with sudden illumination. 'Why, if we're going up country and are likely to see some of your friends, I'll put on my blue dress, and you must wear your watch; I am not going at all if you mean to wear the big hat.'

'Now you're behavin' pretty,' responded Mrs Todd, with a gay toss of her head and a cheerful smile, as she came across the room, bringing a saucerful of wild raspberries, a pretty piece of salvage from supper-time. 'I was cast down when I see you come to breakfast. I didn't think 't was just what you'd select to wear to the reunion, where you're goin' to meet everybody.'

'What reunion do you mean?' I asked, not without amazement. 'Not the Bowden Family's? I thought that was going to take place in September.'

'To-day's the day. They sent word the middle o' the week. I thought you might have heard of it. Yes, they changed the day. I been thinkin' we'd talk it over, but you never can tell beforehand how it's goin' to be, and 't ain't worth while to wear a day all out before it comes.' Mrs Todd gave no place to the pleasures of anticipation, but she spoke like the oracle that she was. 'I wish mother was here to go,' she continued sadly. 'I did look for her last night, and I couldn't keep back the tears when the dark really fell and she wa'n't here, she does so enjoy a great occasion. If William had a mite o' snap an' ambition, he'd take the lead at such a time. Mother likes variety, and there ain't but a few nice opportunities 'round here, an' them she has to miss 'less she contrives to get ashore to me. I do re'lly hate to go to the reunion without mother, an' 't is a beautiful day; everybody'll be asking where she is. Once she'd have got here anyway. Poor mother's beginnin' to feel her age.'

'Why, there's your mother now!' I exclaimed with joy, I was so glad to see the dear old soul again. 'I hear her voice at the gate.' But Mrs Todd was out of the door before me.

There, sure enough, stood Mrs Blackett, who must have left Green Island before daylight. She had climbed the steep road from the water-side so eagerly that she was out of breath, and was standing by the garden fence to rest. She held an old-fashioned brown wicker cap-basket in her hand, as if visiting were a thing of every day, and looked up at us as pleased and triumphant as a child.

'Oh, what a poor, plain garden! Hardly a flower in it except your bush o' balm!' she said. 'But you do keep your garden neat, Almiry. Are you both well, an' goin' up country with me?' She came a step or two closer to meet us, with quaint politeness and quite as delightful as if she were at home. She dropped a quick little curtsey before Mrs Todd.

'There, mother, what a girl you be! I am so pleased! I was just bewailin' you,' said the daughter, with unwonted feeling. 'I was just bewailin' you, I was so disappointed, an' I kep' myself awake a good piece o' the night scoldin' poor William. I watched for the boat till I was ready to shed tears yisterday, and when 't was comin' dark I kep' making errands out to the gate an' down the road to see if you wa'n't in the doldrums somewhere down the bay.'

'There was a head wind, as you know,' said Mrs Blackett, giving me the cap-basket, and holding my hand affectionately as we walked up the clean-swept path to the door. 'I was partly ready to come, but dear William said I should be all tired out and might get cold, havin' to beat all the way in. So we give it up, and set down and spent the evenin' together. It was a little rough and windy outside, and I guess 't was better judgment; we went to bed very early and made a good start just at daylight. It's been a lovely mornin' on the water. William thought he'd better fetch across beyond Bird Rocks, rowin' the greater part o' the way; then we sailed from there right over to the Landin', makin' only one tack. William'll be in again for me to-morrow, so I can come back here an' rest me over night, an' go to meetin' to-morrow, and have a nice, good visit.'

'She was just havin' her breakfast,' said Mrs Todd, who had listened eagerly to the long explanation without a word of disapproval, while her face shone more and more with joy. 'You just sit right down an' have a cup of tea and rest you while we make our preparations. Oh, I am so gratified to think you've come! Yes, she was just havin' her breakfast, and we were speakin' of you. Where's William?'

'He went right back; he said he expected some schooners in about noon after bait, but he'll come an' have his dinner with us to-morrow, unless it rains; then next day. I laid his best things out all ready,' explained Mrs Blackett, a little anxiously. 'This wind will serve him nice all the way home. Yes, I will take a cup of tea, dear,—a cup of tea is always good; and then I'll rest a minute and be all ready to start.'

'I do feel condemned for havin' such hard thoughts o' William,' openly confessed Mrs Todd. She stood before us so large and serious that we both laughed and could not find it in our hearts to convict so rueful a culprit. 'He shall have a good dinner to-morrow, if it can be got, and I shall be real glad to see William,' the confession ended handsomely, while Mrs Blackett smiled approval and made haste to praise the tea. Then I hurried away to make sure of the grocery wagon. Whatever might be the good of the reunion, I was going to have the pleasure and delight of a day in Mrs Blackett's company, not to speak of Mrs Todd's.

The early morning breeze was still blowing, and the warm, sunshiny air was of some ethereal northern sort, with a cool freshness as if it came over new-fallen snow. The world was filled with a fragrance of fir-balsam and the faintest flavor of seaweed from the ledges, bare and brown at low tide in the little harbor. It was so still and so early that the village was but half awake. I could hear no voices but those of the birds, small and great,—the constant song sparrows, the clink of a yellow-hammer over in the woods, and the far conversation of some deliberate crows. I saw William Blackett's escaping sail already far from land, and Captain Littlepage was sitting behind his closed window as I passed by, watching for some one who never came. I tried to speak to him, but he did not see me. There was a patient look on the old man's face, as if the world were a great mistake and he had nobody with whom to speak his own language or find companionship.

XVII

A COUNTRY ROAD

WHATEVER doubts and anxieties I may have had about the inconvenience of the Beggs' high wagon for a person of Mrs Blackett's age and shortness, they were happily overcome by the aid of a chair and her own valiant spirit. Mrs Todd bestowed great care upon seating us as if we were taking passage by boat, but she finally pronounced that we were properly trimmed. When we had gone only a little way up the hill she remembered that she had left the house door wide open, though the large key was safe in her pocket. I offered to run back, but my offer was met with lofty scorn, and we lightly dismissed the matter from our minds, until two or three miles further on we met the doctor, and Mrs Todd asked him to stop and ask her nearest neighbor to step over and close the door if the dust seemed to blow in the afternoon.

'She'll be there in her kitchen; she'll hear you the minute you call; 't won't give you no delay,' said Mrs Todd to the doctor. 'Yes, Mis' Dennett's right there, with the windows all open. It isn't as if my fore door opened right on the road, anyway.' At which proof of composure Mrs Blackett smiled wisely at me.

The doctor seemed delighted to see our guest; they were evidently the warmest friends, and I saw a look of affectionate confidence in their eyes. The good man left his carriage to speak to us, but as he took Mrs Blackett's hand he held it a moment, and, as if merely from force of habit, felt her pulse as they talked; then to my delight he gave the firm old wrist a commending pat.

'You're wearing well: good for another ten years at this rate,' he assured her cheerfully, and she smiled back. 'I like to keep a strict account of my old stand-bys,' and he turned to me. 'Don't you let Mrs Todd overdo to-day,—old folks like her are apt to be thoughtless;' and then we all laughed, and, parting, went our ways gayly.

'I suppose he puts up with your rivalry the same as ever?'

asked Mrs Blackett. 'You and he are as friendly as ever, I see, Almiry,' and Almira sagely nodded.

'He's got too many long routes now to stop to 'tend to all his door patients,' she said, 'especially them that takes pleasure in talkin' themselves over. The doctor and me have got to be kind of partners; he's gone a good deal, far an' wide. Looked tired, didn't he? I shall have to advise with him an' get him off for a good rest. He'll take the big boat from Rockland an' go off up to Boston an' mouse round among the other doctors, once in two or three years, and come home fresh as a boy. I guess they think consider'ble of him up there.' Mrs Todd shook the reins and reached determinedly for the whip, as if she were compelling public opinion.

Whatever energy and spirit the white horse had to begin with were soon exhausted by the steep hills and his discernment of a long expedition ahead. We toiled slowly along. Mrs Blackett and I sat together, and Mrs Todd sat alone in front with much majesty and the large basket of provisions. Part of the way the road was shaded by thick woods, but we also passed one farmhouse after another on the high uplands, which we all three regarded with deep interest, the house itself and the barns and garden-spots and poultry all having to suffer an inspection of the shrewdest sort. This was a highway quite new to me; in fact, most of my journeys with Mrs Todd had been made afoot and between the roads, in open pasturelands. My friends stopped several times for brief dooryard visits, and made so many promises of stopping again on the way home that I began to wonder how long the expedition would last. I had often noticed how warmly Mrs Todd was greeted by her friends, but it was hardly to be compared to the feeling now shown toward Mrs Blackett. A look of delight came to the faces of those who recognized the plain, dear old figure beside me; one revelation after another was made of the constant interest and intercourse that had linked the far island and these scattered farms into a golden chain of love and dependence.

'Now, we mustn't stop again if we can help it,' insisted Mrs Todd at last. 'You'll get tired, mother, and you'll think the less o' reunions. We can visit along here any day. There, if they ain't frying doughnuts in this next house, too! These are new folks, you know, from over St George way; they took this old Talcot farm last year. 'T is the best water on the road, and the check-rein's come undone—yes, we'd best delay a little and water the horse.'

We stopped, and seeing a party of pleasure-seekers in holiday attire, the thin, anxious mistress of the farmhouse came out with wistful sympathy to hear what news we might have to give. Mrs Blackett first spied her at the half-closed door, and asked with such cheerful directness if we were trespassing that, after a few words, she went back to her kitchen and reappeared with a plateful of doughnuts.

'Entertainment for man and beast,'* announced Mrs Todd with satisfaction. 'Why, we've perceived there was new doughnuts all along the road, but you're the first that has treated us.'

Our new acquaintance flushed with pleasure, but said nothing.

'They're very nice; you've had good luck with 'em,' pronounced Mrs Todd. 'Yes, we've observed there was doughnuts all the way along; if one house is frying all the rest is; 't is so with a great many things.'

'I don't suppose likely you're goin' up to the Bowden reunion?' asked the hostess as the white horse lifted his head and we were saying good-by.

'Why, yes,' said Mrs Blackett and Mrs Todd and I, all together.

'I am connected with the family. Yes, I expect to be there this afternoon. I've been lookin' forward to it,' she told us eagerly.

'We shall see you there. Come and sit with us if it's convenient,' said dear Mrs Blackett, and we drove away.

'I wonder who she was before she was married?' said Mrs Todd, who was usually unerring in matters of genealogy. 'She must have been one of that remote branch that lived down beyond Thomaston. We can find out this afternoon. I expect that the families'll march together, or be sorted out some way. I'm willing to own a relation that has such proper ideas of doughnuts.'

'I seem to see the family looks,' said Mrs Blackett. 'I wish we'd asked her name. She's a stranger, and I want to help make it pleasant for all such.'

'She resembles Cousin Pa'lina Bowden about the forehead,' said Mrs Todd with decision.

We had just passed a piece of woodland that shaded the road, and come out to some open fields beyond, when Mrs Todd suddenly reined in the horse as if somebody had stood on the roadside and stopped her. She even gave that quick reassuring nod of her head which was usually made to answer for a bow, but I discovered that she was looking eagerly at a tall ash-tree that grew just inside the field fence.

'I thought 't was goin' to do well,' she said complacently as we went on again. 'Last time I was up this way that tree was kind of drooping and discouraged. Grown trees act that way sometimes, same 's folks; then they'll put right to it and strike their roots off into new ground and start all over again with real good courage. Ash-trees is very likely to have poor spells; they ain't got the resolution of other trees.'

I listened hopefully for more; it was this peculiar wisdom that made one value Mrs Todd's pleasant company.

'There's sometimes a good hearty tree growin' right out of the bare rock, out o' some crack that just holds the roots;' she went on to say, 'right on the pitch o' one o' them bare stony hills where you can't seem to see a wheel-barrowful o' good earth in a place, but that tree'll keep a green top in the driest summer. You lay your ear down to the ground an' you'll hear a little stream runnin'. Every such tree has got its own livin' spring;* there's folks made to match 'em.'

I could not help turning to look at Mrs Blackett, close beside me. Her hands were clasped placidly in their thin black woolen gloves, and she was looking at the flowery wayside as we went slowly along, with a pleased, expectant smile. I do not think she had heard a word about the trees.

'I just saw a nice plant o' elecampane growin' back there,' she said presently to her daughter.

'I haven't got my mind on herbs to-day,' responded Mrs Todd, in the most matter-of-fact way. 'I'm bent on seeing folks,' and she shook the reins again.

I for one had no wish to hurry, it was so pleasant in the shady roads. The woods stood close to the road on the right; on the left were narrow fields and pastures where there were as many acres of spruces and pines as there were acres of bay and juniper and huckleberry, with a little turf between. When I thought we were in the heart of the inland country, we reached the top of a hill, and suddenly there lay spread out before us a wonderful great view of well-cleared fields that swept down to the wide water of a bay. Beyond this were distant shores like another country in the midday haze which half hid the hills beyond, and the far-away pale blue mountains on the northern horizon. There was a schooner with all sails set coming down the bay from a white village that was sprinkled on the shore, and there were many sailboats flitting about. It was a noble landscape, and my eyes, which had grown used to the narrow inspection of a shaded roadside, could hardly take it in.

'Why, it's the upper bay,' said Mrs Todd. 'You can see 'way over into the town of Fessenden. Those farms 'way over there are all in Fessenden. Mother used to have a sister that lived up that shore. If we started as early 's we could on a summer mornin', we couldn't get to her place from Green Island till late afternoon, even with a fair, steady breeze, and you had to strike the time just right so as to fetch up 'long o' the tide and land near the flood. 'T was ticklish business, an' we didn't visit back an' forth as much as mother desired. You have to go 'way down the co'st to Cold Spring Light an' round that long point,—up here's what they call the Back Shore.'

'No, we were 'most always separated, my dear sister and me, after the first year she was married,' said Mrs Blackett. 'We had our little families an' plenty o' cares. We were always lookin' forward to the time we could see each other more. Now and then she'd get out to the island for a few days while her husband'd go fishin'; and once he stopped with her an' two children, and made him some flakes right there and cured all his fish for winter. We did have a beautiful time together, sister an' me; she used to look back to it long 's she lived.'

'I do love to look over there where she used to live,' Mrs

Blackett went on as we began to go down the hill. 'It seems as if she must still be there, though she's long been gone. She loved their farm,—she didn't see how I got so used to our island; but somehow I was always happy from the first.'

'Yes, it's very dull to me up among those slow farms,' declared Mrs Todd. 'The snow troubles 'em in winter. They're all besieged by winter, as you may say; 't is far better by the shore than up among such places. I never thought I should like to live up country.'

'Why, just see the carriages ahead of us on the next rise!' exclaimed Mrs Blackett. 'There's going to be a great gathering, don't you believe there is, Almiry? It hasn't seemed up to now as if anybody was going but us. An' 't is such a beautiful day, with yesterday cool and pleasant to work an' get ready, I shouldn't wonder if everybody was there, even the slow ones like Phebe Ann Brock.'

Mrs Blackett's eyes were bright with excitement, and even Mrs Todd showed remarkable enthusiasm. She hurried the horse and caught up with the holiday-makers ahead. 'There's all the Dep'fords goin', six in the wagon,' she told us joyfully; 'an' Mis' Alva Tilley's folks are now risin' the hill in their new carryall.'

Mrs Blackett pulled at the neat bow of her black bonnet-strings, and tied them again with careful precision. 'I believe your bonnet's on a little bit sideways, dear,' she advised Mrs Todd as if she were a child; but Mrs Todd was too much occupied to pay proper heed. We began to feel a new sense of gayety and of taking part in the great occasion as we joined the little train.

XVIII

THE BOWDEN REUNION

IT is very rare in country life, where high days and holidays
are few, that any occasion of general interest proves to be less
than great. Such is the hidden fire of enthusiasm in the New
England nature that, once given an outlet, it shines forth with
almost volcanic light and heat. In quiet neighborhoods such
inward force does not waste itself upon those petty excite-
ments of every day that belong to cities, but when, at long
intervals, the altars to patriotism, to friendship, to the ties of
kindred, are reared in our familiar fields, then the fires glow,
the flames come up as if from the inexhaustible burning
heart of the earth; the primal fires break through the granite
dust in which our souls are set. Each heart is warm and every
face shines with the ancient light. Such a day as this has
transfiguring powers, and easily makes friends of those who
have been cold-hearted, and gives to those who are dumb
their chance to speak, and lends some beauty to the plainest
face.

'Oh, I expect I shall meet friends to-day that I haven't seen
in a long while,' said Mrs Blackett with deep satisfaction. ''T
will bring out a good many of the old folks, 't is such a lovely
day. I'm always glad not to have them disappointed.'

'I guess likely the best of 'em'll be there,' answered
Mrs Todd with gentle humor, stealing a glance at me.
'There's one thing certain: there's nothing takes in this whole
neighborhood like anything related to the Bowdens. Yes, I do
feel that when you call upon the Bowdens you may expect
most families to rise up between the Landing and the far end
of the Back Cove. Those that aren't kin by blood are kin by
marriage.'

'There used to be an old story goin' about when I was a
girl,' said Mrs Blackett, with much amusement. 'There was a
great many more Bowdens then than there are now, and the
folks was all setting in meeting a dreadful hot Sunday afternoon, and a scatter-witted little bound girl* came running to

the meetin'-house door all out o' breath from somewheres in the neighborhood. "Mis' Bowden, Mis' Bowden!" says she. "Your baby's in a fit!" They used to tell that the whole congregation was up on its feet in a minute and right out into the aisles. All the Mis' Bowdens was setting right out for home; the minister stood there in the pulpit tryin' to keep sober, an' all at once he burst right out laughin'. He was a very nice man, they said, and he said he'd better give 'em the benediction, and they could hear the sermon next Sunday, so he kept it over. My mother was there, and she thought certain 't was me.'

'None of our family was ever subject to fits,' interrupted Mrs Todd severely. 'No, we never had fits, none of us, and 't was lucky we didn't 'way out there to Green Island. Now these folks right in front: dear sakes knows the bunches o' soothing catnip an' yarrow I've had to favor old Mis' Evins with dryin'! You can see it right in their expressions, all them Evins folks. There, just you look up to the cross-roads, mother,' she suddenly exclaimed. 'See all the teams ahead of us. And oh, look down on the bay; yes, look down on the bay! See what a sight o' boats, all headin' for the Bowden place cove!'

'Oh, ain't it beautiful!' said Mrs Blackett, with all the delight of a girl. She stood up in the high wagon to see everything, and when she sat down again she took fast hold of my hand.

'Hadn't you better urge the horse a little, Almiry?' she asked. 'He's had it easy as we came along, and he can rest when we get there. The others are some little ways ahead, and I don't want to lose a minute.'

We watched the boats drop their sails one by one in the cove as we drove along the high land. The old Bowden house stood, low-storied and broad-roofed, in its green fields as if it were a motherly brown hen waiting for the flock that came straying toward it from every direction. The first Bowden settler had made his home there, and it was still the Bowden farm; five generations of sailors and farmers and soldiers had been its children. And presently Mrs Blackett showed me the stone-walled burying-ground that stood like a little fort on a knoll overlooking the bay, but, as she said, there were plenty of scattered Bowdens who were not laid there,—some lost at sea, and some out West, and some who died in the war; most of the home graves were those of women.

We could see now that there were different footpaths from along shore and across country. In all these there were straggling processions walking in single file, like old illustrations of the Pilgrim's Progress.* There was a crowd about the house as if huge bees were swarming in the lilac bushes. Beyond the fields and cove a higher point of land ran out into the bay, covered with woods which must have kept away much of the northwest wind in winter. Now there was a pleasant look of shade and shelter there for the great family meeting.

We hurried on our way, beginning to feel as if we were very late, and it was a great satisfaction at last to turn out of the stony highroad into a green lane shaded with old apple-trees. Mrs Todd encouraged the horse until he fairly pranced with gayety as we drove round to the front of the house on the soft turf. There was an instant cry of rejoicing, and two or three persons ran toward us from the busy group.

'Why, dear Mis' Blackett!—here's Mis' Blackett!' I heard them say, as if it were pleasure enough for one day to have a sight of her. Mrs Todd turned to me with a lovely look of triumph and self-forgetfulness. An elderly man who wore the look of a prosperous sea-captain put up both arms and lifted Mrs Blackett down from the high wagon like a child, and kissed her with hearty affection. 'I was master afraid* she wouldn't be here,' he said, looking at Mrs Todd with a face like a happy sunburnt schoolboy, while everybody crowded round to give their welcome.

'Mother's always the queen,' said Mrs Todd. 'Yes, they'll all make everything of mother; she'll have a lovely time to-day. I wouldn't have had her miss it, and there won't be a thing she'll ever regret, except to mourn because William wa'n't here.'

Mrs Blackett having been properly escorted to the house, Mrs Todd received her own full share of honor, and some of the men, with a simple kindness that was the soul of chivalry, waited upon us and our baskets and led away the white horse. I already knew some of Mrs Todd's friends and kindred, and

felt like an adopted Bowden in this happy moment. It seemed
to be enough for any one to have arrived by the same conveyance as Mrs Blackett, who presently had her court inside the
house, while Mrs Todd, large, hospitable, and preëminent,
was the centre of a rapidly increasing crowd about the lilac
bushes. Small companies were continually coming up the
long green slope from the water, and nearly all the boats had
come to shore. I counted three or four that were baffled by
the light breeze, but before long all the Bowdens, small and
great, seemed to have assembled, and we started to go up to
the grove across the field.

Out of the chattering crowd of noisy children, and large-
waisted women whose best black dresses fell straight to the
ground in generous folds, and sunburnt men who looked as
serious as if it were town-meeting day, there suddenly came
silence and order. I saw the straight, soldierly little figure of a
man who bore a fine resemblance to Mrs Blackett, and who
appeared to marshal us with perfect ease. He was imperative
enough, but with a grand military sort of courtesy, and bore
himself with solemn dignity of importance. We were sorted
out according to some clear design of his own, and stood as
speechless as a troop to await his orders. Even the children
were ready to march together, a pretty flock, and at the last
moment Mrs Blackett and a few distinguished companions,
the ministers and those who were very old, came out of the
house together and took their places. We ranked by fours,
and even then we made a long procession.

There was a wide path mowed for us across the field, and,
as we moved along, the birds flew up out of the thick second
crop of clover, and the bees hummed as if it still were June.
There was a flashing of white gulls over the water where the
fleet of boats rode the low waves together in the cove, swaying
their small masts as if they kept time to our steps. The plash of
the water could be heard faintly, yet still be heard; we might
have been a company of ancient Greeks going to celebrate a
victory, or to worship the god of harvests* in the grove above.
It was strangely moving to see this and to make part of it. The
sky, the sea, have watched poor humanity at its rites so long;
we were no more a New England family celebrating its own

existence and simple progress; we carried the tokens and inheritance of all such households from which this had descended, and were only the latest of our line. We possessed the instincts of a far, forgotten childhood; I found myself thinking that we ought to be carrying green branches and singing as we went. So we came to the thick shaded grove still silent, and were set in our places by the straight trees that swayed together and let sunshine through here and there like a single golden leaf that flickered down, vanishing in the cool shade.

The grove was so large that the great family looked far smaller than it had in the open field; there was a thick growth of dark pines and firs with an occasional maple or oak that gave a gleam of color like a bright window in the great roof. On three sides we could see the water, shining behind the tree-trunks, and feel the cool salt breeze that began to come up with the tide just as the day reached its highest point of heat. We could see the green sunlit field we had just crossed as if we looked out at it from a dark room, and the old house and its lilacs standing placidly in the sun, and the great barn with a stockade of carriages from which two or three caretaking men who had lingered were coming across the field together. Mrs Todd had taken off her warm gloves and looked the picture of content.

'There!' she exclaimed. 'I've always meant to have you see this place, but I never looked for such a beautiful opportunity—weather an' occasion both made to match. Yes, it suits me: I don't ask no more. I want to know if you saw mother walkin' at the head! It choked me right up to see mother at the head, walkin' with the ministers,' and Mrs Todd turned away to hide the feelings she could not instantly control.

'Who was the marshal?' I hastened to ask. 'Was he an old soldier?'

'Don't he do well?' answered Mrs Todd with satisfaction.

'He don't often have such a chance to show off his gifts,' said Mrs Caplin, a friend from the Landing who had joined us. 'That's Sant Bowden; he always takes the lead, such days. Good for nothing else most o' his time; trouble is, he'—

I turned with interest to hear the worst. Mrs Caplin's tone was both zealous and impressive.

'Stim'lates,' she explained scornfully.

'No, Santin never was in the war,' said Mrs Todd with lofty indifference. 'It was a cause of real distress to him. He kep' enlistin', and traveled far an' wide about here, an' even took the bo't and went to Boston to volunteer; but he ain't a sound man, an' they wouldn't have him. They say he knows all their tactics, an' can tell all about the battle o' Waterloo* well 's he can Bunker Hill.* I told him once the country'd lost a great general, an' I meant it, too.'

'I expect you're near right,' said Mrs Caplin, a little crestfallen and apologetic.

'I be right,' insisted Mrs Todd with much amiability. ''T was most too bad to cramp him down to his peaceful trade, but he's a most excellent shoemaker at his best, an' he always says it's a trade that gives him time to think an' plan his manœuvres. Over to the Port they always invite him to march Decoration Day,* same as the rest, an' he does look noble; he comes of soldier stock.'

I had been noticing with great interest the curiously French type of face which prevailed in this rustic company. I had said to myself before that Mrs Blackett was plainly of French descent, in both her appearance and her charming gifts, but this is not surprising when one has learned how large a proportion of the early settlers on this northern coast of New England were of Huguenot blood, and that it is the Norman Englishman, not the Saxon, who goes adventuring to a new world.

'They used to say in old times,' said Mrs Todd modestly, 'that our family came of very high folks in France, and one of 'em was a great general in some o' the old wars. I sometimes think that Santin's ability has come 'way down from then. 'T ain't nothin' he's ever acquired; 't was born in him. I don't know 's he ever saw a fine parade, or met with those that studied up such things. He's figured it all out an' got his papers so he knows how to aim a cannon right for William's fish-house five miles out on Green Island, or up there on

Burnt Island where the signal is. He had it all over to me one day, an' I tried hard to appear interested. His life's all in it, but he will have those poor gloomy spells come over him now an' then, an' then he has to drink.'

Mrs Caplin gave a heavy sigh.

'There's a great many such strayaway folks, just as there is plants,' continued Mrs Todd, who was nothing if not botanical. 'I know of just one sprig of laurel that grows over back here in a wild spot, an' I never could hear of no other on this coast. I had a large bunch brought me once from Massachusetts way, so I know it. This piece grows in an open spot where you'd think 't would do well, but its sort o' poor-lookin'. I've visited it time an' again, just to notice its poor blooms. 'T is a real Sant Bowden, out of its own place.'

Mrs Caplin looked bewildered and blank. 'Well, all I know is, last year he worked out some kind of a plan so 's to parade the county conference in platoons, and got 'em all flustered up tryin' to sense his ideas of a holler square,' she burst forth. 'They was holler enough anyway after ridin' 'way down from up country into the salt air, and they'd been treated to a sermon on faith an' works from old Fayther Harlow that never knows when to cease. 'T wa'n't no time for tactics then,—they wa'n't a-thinkin' of the church military.* Sant, he couldn't do nothin' with 'em. All he thinks of, when he sees a crowd, is how to march 'em. 'T is all very well when he don't 'tempt too much. He never did act like other folks.'

'Ain't I just been maintainin' that he ain't like 'em?' urged Mrs Todd decidedly. 'Strange folks has got to have strange ways, for what I see.'

'Somebody observed once that you could pick out the likeness of 'most every sort of a foreigner when you looked about you in our parish,' said Sister Caplin, her face brightening with sudden illumination. 'I didn't see the bearin' of it then quite so plain. I always did think Mari' Harris resembled a Chinee.'

'Mari' Harris was pretty as a child, I remember,' said the pleasant voice of Mrs Blackett, who, after receiving the affectionate greetings of nearly the whole company, came to join us,—to see, as she insisted, that we were out of mischief.

'Yes, Mari' was one o' them pretty little lambs that make dreadful homely old sheep,' replied Mrs Todd with energy. 'Cap'n Littlepage never'd look so disconsolate if she was any sort of a proper person to direct things. She might divert him; yes, she might divert the old gentleman, an' let him think he had his own way, 'stead o' arguing everything down to the bare bone. 'T wouldn't hurt her to sit down an' hear his great stories once in a while.'

'The stories are very interesting,' I ventured to say.

'Yes, you always catch yourself a-thinkin' what if they was all true, and he had the right of it,' answered Mrs Todd. 'He's a good sight better company, though dreamy, than such sordid creatur's as Mari' Harris.'

'Live and let live,' said dear old Mrs Blackett gently. 'I haven't seen the captain for a good while, now that I ain't so constant to meetin',' she added wistfully. 'We always have known each other.'

'Why, if it is a good pleasant day to-morrow, I'll get William to call an' invite the capt'in to dinner. William'll be in early so 's to pass up the street without meetin' anybody.'

'There, they're callin' out it's time to set the tables,' said Mrs Caplin, with great excitement.

'Here's Cousin Sarah Jane Blackett! Well, I am pleased, certain!' exclaimed Mrs Todd, with unaffected delight; and these kindred spirits met and parted with the promise of a good talk later on. After this there was no more time for conversation until we were seated in order at the long tables.

'I'm one that always dreads seeing some o' the folks that I don't like, at such a time as this,' announced Mrs Todd privately to me after a season of reflection. We were just waiting for the feast to begin. 'You wouldn't think such a great creatur' 's I be could feel all over pins an' needles. I remember, the day I promised to Nathan, how it come over me, just's I was feelin' happy's I could, that I'd got to have an own cousin o' his for my near relation all the rest o' my life, an' it seemed as if die I should. Poor Nathan saw somethin' had crossed me,—he had very nice feelings,—and when he asked me what 't was, I told him. "I never could like her myself," said he. "You sha'n't be bothered, dear," he says; an' 't was one o' the things that made me set a good deal by Nathan, he didn't make a habit of always opposin', like some men. "Yes," says I, "but think o' Thanksgivin' times an' funerals; she's our relation, an' we've got to own her." Young folks don't think o' those things. There she goes now, do let's pray her by!' said Mrs Todd, with an alarming transition from general opinions to particular animosities. 'I hate her just the same as I always did; but she's got on a real pretty dress. I do try to remember that she's Nathan's cousin. Oh dear, well; she's gone by after all, an' ain't seen me. I expected she'd come pleasantin' round just to show off an' say afterwards she was acquainted.'

This was so different from Mrs Todd's usual largeness of mind that I had a moment's uneasiness; but the cloud passed quickly over her spirit, and was gone with the offender.

There never was a more generous out-of-door feast along the coast than the Bowden family set forth that day. To call it a picnic would make it seem trivial. The great tables were edged with pretty oak-leaf trimming, which the boys and girls made. We brought flowers from the fence-thickets of the great field; and out of the disorder of flowers and provisions suddenly appeared as orderly a scheme for the feast as the marshal had shaped for the procession. I began to respect the Bowdens for their inheritance of good taste and skill and a certain pleasing gift of formality. Something made them do all these things in a finer way than most country people would have done them. As I looked up and down the tables there was a good cheer, a grave soberness that shone with pleasure, a humble dignity of bearing. There were some who should have sat below the salt* for lack of this good breeding; but they were not many. So, I said to myself, their ancestors may have sat in the great hall of some old French house in the Middle Ages, when battles and sieges and processions and feasts were familiar things. The ministers and Mrs Blackett, with a few of their rank and age, were put in places of honor, and for once that I looked any other way I looked twice at Mrs Blackett's face, serene and mindful of privilege and responsibility, the mistress by simple fitness of this great day.

Mrs Todd looked up at the roof of green trees, and then carefully surveyed the company. 'I see 'em better now they're all settin' down,' she said with satisfaction. 'There's old Mr Gilbraith and his sister. I wish they were settin' with us; they're not among folks they can parley with, an' they look disappointed.'

As the feast went on, the spirits of my companion steadily rose. The excitement of an unexpectedly great occasion was a subtle stimulant to her disposition, and I could see that sometimes when Mrs Todd had seemed limited and heavily domestic, she had simply grown sluggish for lack of proper surroundings. She was not so much reminiscent now as expectant, and as alert and gay as a girl. We who were her neighbors were full of gayety, which was but the reflected light from her beaming countenance. It was not the first time that I was full of wonder at the waste of human ability in this world, as a botanist wonders at the wastefulness of nature, the thousand seeds that die, the unused provision of every sort. The reserve force of society grows more and more amazing to one's thought. More than one face among the Bowdens showed that only opportunity and stimulus were lacking,—a narrow set of circumstances had caged a fine able character and held it captive. One sees exactly the same types in a country gathering as in the most brilliant city company. You are safe to be understood if the spirit of your speech is the same for one neighbor as for the other.

XIX

THE FEAST'S END

THE feast was a noble feast, as has already been said. There was an elegant ingenuity displayed in the form of pies which delighted my heart. Once acknowledge that an American pie is far to be preferred to its humble ancestor, the English tart, and it is joyful to be reassured at a Bowden reunion that invention has not yet failed. Beside a delightful variety of material, the decorations went beyond all my former experience; dates and names were wrought in lines of pastry and frosting on the tops. There was even more elaborate reading matter on an excellent early-apple pie which we began to share and eat, precept upon precept. Mrs Todd helped me generously to the whole word *Bowden*, and consumed *Reunion* herself, save an undecipherable fragment; but the most renowned essay in cookery on the tables was a model of the old Bowden house made of durable gingerbread, with all the windows and doors in the right places, and sprigs of genuine lilac set at the front. It must have been baked in sections, in one of the last of the great brick ovens, and fastened together on the morning of the day. There was a general sigh when this fell into ruin at the feast's end, and it was shared by a great part of the assembly, not without seriousness, and as if it were a pledge and token of loyalty. I met the maker of the gingerbread house, which had called up lively remembrances of a childish story. She had the gleaming eye of an enthusiast and a look of high ideals.

'I could just as well have made it all of frosted cake,' she said, 'but 't wouldn't have been the right shade; the old house, as you observe, was never painted, and I concluded that plain gingerbread would represent it best. It wasn't all I expected it would be,' she said sadly, as many an artist had said before her of his work.

There were speeches by the ministers; and there proved to be a historian among the Bowdens, who gave some fine anecdotes of the family history; and then appeared a poetess,

whom Mrs Todd regarded with wistful compassion and indul-
gence, and when the long faded garland of verses came to an
appealing end, she turned to me with words of praise.

'Sounded pretty,' said the generous listener. 'Yes, I thought
she did very well. We went to school together, an' Mary Anna
had a very hard time; trouble was, her mother thought she'd
given birth to a genius, an' Mary Anna's come to believe it
herself. There, I don't know what we should have done without her; there ain't nobody else that can write poetry between
here and 'way up towards Rockland; it adds a great deal at
such a time. When she speaks o' those that are gone, she feels
it all, and so does everybody else, but she harps too much. I'd
laid half of that away for next time, if I was Mary Anna. There
comes mother to speak to her, an' old Mr Gilbraith's sister;
now she'll be heartened right up. Mother'll say just the right
thing.'

The leave-takings were as affecting as the meetings of these
old friends had been. There were enough young persons at
the reunion; but it is the old who really value such opportunities; as for the young, it is the habit of every day to meet
their comrades,—the time of separation has not come. To see
the joy with which these elder kinsfolk and acquaintances had
looked in one another's faces, and the lingering touch of
their friendly hands; to see these affectionate meetings and
then the reluctant partings, gave one a new idea of the isola-
tion in which it was possible to live in that after all thinly
settled region. They did not expect to see one another again
very soon; the steady, hard work on the farms, the difficulty of
getting from place to place, especially in winter when boats
were laid up, gave double value to any occasion which could
bring a large number of families together. Even funerals in
this country of the pointed firs were not without their social
advantages and satisfactions. I heard the words 'next summer'
repeated many times, though summer was still ours and all
the leaves were green.

The boats began to put out from shore, and the wagons to
drive away. Mrs Blackett took me into the old house when we
came back from the grove: it was her father's birthplace and
early home, and she had spent much of her own childhood

there with her grandmother. She spoke of those days as if they had but lately passed; in fact, I could imagine that the house looked almost exactly the same to her. I could see the brown rafters of the unfinished roof as I looked up the steep staircase, though the best room was as handsome with its good wainscoting and touch of ornament on the cornice as any old room of its day in a town.

Some of the guests who came from a distance were still sitting in the best room when we went in to take leave of the master and mistress of the house. We all said eagerly what a pleasant day it had been, and how swiftly the time had passed. Perhaps it is the great national anniversaries* which our country has lately kept, and the soldiers' meetings that take place everywhere, which have made reunions of every sort the fashion. This one, at least, had been very interesting. I fancied that old feuds had been overlooked, and the old saying that blood is thicker than water had again proved itself true, though from the variety of names one argued a certain adulteration of the Bowden traits and belongings. Clannishness is an instinct of the heart,—it is more than a birthright, or a custom; and lesser rights were forgotten in the claim to a common inheritance.

We were among the very last to return to our proper lives and lodgings. I came near to feeling like a true Bowden, and parted from certain new friends as if they were old friends; we were rich with the treasure of a new remembrance.

At last we were in the high wagon again; the old white horse had been well fed in the Bowden barn, and we drove away and soon began to climb the long hill toward the wooded ridge. The road was new to me, as roads always are, going back. Most of our companions had been full of anxious thoughts of home,—of the cows, or of young children likely to fall into disaster,—but we had no reasons for haste, and drove slowly along, talking and resting by the way. Mrs Todd said once that she really hoped her front door had been shut on account of the dust blowing in, but added that nothing made any weight on her mind except not to forget to turn a few late mullein leaves that were drying on a newspaper in the little loft. Mrs Blackett and I gave our word of honor that we would remind her of this heavy responsibility. The way seemed short, we had so much to talk about. We climbed hills where we could see the great bay and the islands, and then went down into shady valleys where the air began to feel like evening, cool and damp with a fragrance of wet ferns. Mrs Todd alighted once or twice, refusing all assistance in securing some boughs of a rare shrub which she valued for its bark, though she proved incommunicative as to her reasons. We passed the house where we had been so kindly entertained with doughnuts earlier in the day, and found it closed and deserted, which was a disappointment.

'They must have stopped to tea somewheres and thought they'd finish up the day,' said Mrs Todd. 'Those that enjoyed it best'll want to get right home so 's to think it over.'

'I didn't see the woman there after all, did you?' asked Mrs Blackett as the horse stopped to drink at the trough.

'Oh yes, I spoke with her,' answered Mrs Todd, with but scant interest or approval. 'She ain't a member o' our family.'

'I thought you said she resembled Cousin Pa'lina Bowden about the forehead,' suggested Mrs Blackett.

'Well, she don't,' answered Mrs Todd impatiently. 'I ain't one that's ord'narily mistaken about family likenesses, and she didn't seem to meet with friends, so I went square up to her. "I expect you're a Bowden by your looks," says I. "Yes, I take it you're one o' the Bowdens." "Lor', no," says she. "Dennett was my maiden name, but I married a Bowden for my first husband. I thought I'd come an' just see what was a-goin' on"!'

Mrs Blackett laughed heartily. 'I'm goin' to remember to tell William o' that,' she said. 'There, Almiry, the only thing that's troubled me all this day is to think how William would have enjoyed it. I do so wish William had been there.'

'I sort of wish he had, myself,' said Mrs Todd frankly.

'There wa'n't many old folks there, somehow,' said Mrs Blackett, with a touch of sadness in her voice. 'There ain't so many to come as there used to be, I'm aware, but I expected to see more.'

'I thought they turned out pretty well, when you come to think of it; why, everybody was sayin' so an' feelin' gratified,' answered Mrs Todd hastily with pleasing unconsciousness; then I saw the quick color flash into her cheek, and presently she made some excuse to turn and steal an anxioius look at her mother. Mrs Blackett was smiling and thinking about her happy day, though she began to look a little tired. Neither of my companions was troubled by her burden of years. I hoped in my heart that I might be like them as I lived on into age, and then smiled to think that I too was no longer very young. So we always keep the same hearts, though our outer framework fails and shows the touch of time.

''T was pretty when they sang the hymn, wasn't it?' asked Mrs Blackett at suppertime, with real enthusiasm. 'There was such a plenty o' men's voices; where I sat it did sound beautiful. I had to stop and listen when they came to the last verse.'

I saw that Mrs Todd's broad shoulders began to shake. 'There was good singers there; yes, there was excellent singers,' she agreed heartily, putting down her teacup, 'but I chanced to drift alongside Mis' Peter Bowden o' Great Bay, an' I couldn't help thinkin' if she was as far out o' town as she was out o' tune, she wouldn't get back in a day.'

ALONG SHORE

ONE day as I went along the shore beyond the old wharves and the newer, high-stepped fabric of the steamer landing, I saw that all the boats were beached, and the slack water period of the early afternoon prevailed. Nothing was going on, not even the most leisurely of occupations, like baiting trawls or mending nets, or repairing lobster pots; the very boats seemed to be taking an afternoon nap in the sun. I could hardly discover a distant sail as I looked seaward, except a weather-beaten lobster smack, which seemed to have been taken for a plaything by the light airs that blew about the bay. It drifted and turned about so aimlessly in the wide reach off Burnt Island, that I suspected there was nobody at the wheel, or that she might have parted her rusty anchor chain while all the crew were asleep.

I watched her for a minute or two; she was the old Miranda, owned by some of the Caplins, and I knew her by an odd shaped patch of newish duck that was set into the peak of her dingy mainsail. Her vagaries offered such an exciting subject for conversation that my heart rejoiced at the sound of a hoarse voice behind me. At that moment, before I had time to answer, I saw something large and shapeless flung from the Miranda's deck that splashed the water high against her black side, and my companion gave a satisfied chuckle. The old lobster smack's sail caught the breeze again at this moment, and she moved off down the bay. Turning, I found old Elijah Tilley, who had come softly out of his dark fish house, as if it were a burrow.

'Boy got kind o' drowsy steerin' of her; Monroe he hove him right overboard; 'wake now fast enough,' explained Mr Tilley, and we laughed together.

I was delighted, for my part, that the vicissitudes and dangers of the Miranda, in a rocky channel, should have given me this opportunity to make acquaintance with an old fisherman to whom I had never spoken. At first he had seemed to be one of those evasive and uncomfortable persons who are so suspicious of you that they make you almost suspicious of yourself. Mr Elijah Tilley appeared to regard a stranger with scornful indifference. You might see him standing on the pebble beach or in a fishhouse doorway, but when you came nearer he was gone. He was one of the small company of elderly, gaunt-shaped great fishermen whom I used to like to see leading up a deep-laden boat by the head, as if it were a horse, from the water's edge to the steep slope of the pebble beach. There were four of these large old men at the Landing, who were the survivors of an earlier and more vigorous generation. There was an alliance and understanding between them, so close that it was apparently speechless. They gave much time to watching one another's boats go out or come in; they lent a ready hand at tending one another's lobster traps in rough weather; they helped to clean the fish, or to sliver porgies for the trawls, as if they were in close partnership; and when a boat came in from deep-sea fishing they were never far out of the way, and hastened to help carry it ashore, two by two, splashing alongside, or holding its steady head, as if it were a willful sea colt. As a matter of fact no boat could help being steady and way-wise under their instant direction and companionship. Abel's boat and Jonathan Bowden's boat were as distinct and experienced personalities as the men themselves, and as inexpressive. Arguments and opinions were unknown to the conversation of these ancient friends; you would as soon have expected to hear small talk in a company of elephants as to hear old Mr Bowden or Elijah Tilley and their two mates waste breath upon any form of trivial gossip. They made brief statements to one another from time to time. As you came to know them you wondered more and more that they should talk at all. Speech seemed to be a light and elegant accomplishment, and their unexpected acquaintance with its arts made them of new value to the listener. You felt almost as if a landmark pine should suddenly address you in regard to the weather, or a lofty-minded old camel make a remark as you stood respectfully near him under the circus tent.

I often wondered a great deal about the inner life and

thought of these self-contained old fishermen; their minds seemed to be fixed upon nature and the elements rather than upon any contrivances of man, like politics or theology. My friend, Captain Bowden, who was the nephew of the eldest of this group, regarded them with deference; but he did not belong to their secret companionship, though he was neither young nor talkative.

'They've gone together ever since they were boys, they know most everything about the sea amon'st them,' he told me once. 'They was always just as you see 'em now since the memory of man.'

These ancient seafarers had houses and lands not outwardly different from other Dunnet Landing dwellings, and two of them were fathers of families, but their true dwelling places were the sea, and the stony beach that edged its familiar shore, and the fishhouses, where much salt brine from the mackerel kits had soaked the very timbers into a state of brown permanence and petrifaction. It had also affected the old fishermen's hard complexions, until one fancied that when Death claimed them it could only be with the aid, not of any slender modern dart, but the good serviceable harpoon of a seventeenth century woodcut.

Elijah Tilley was such an evasive, discouraged-looking person, heavy-headed, and stooping so that one could never look him in the face, that even after his friendly exclamation about Monroe Pennell, the lobster smack's skipper, and the sleepy boy, I did not venture at once to speak again. Mr Tilley was carrying a small haddock in one hand, and presently shifted it to the other hand lest it might touch my skirt. I knew that my company was accepted, and we walked together a little way.

'You mean to have a good supper,' I ventured to say, by way of friendliness.

'Goin' to have this 'ere haddock an' some o' my good baked potatoes; must eat to live,' responded my companion with great pleasantness and open approval. I found that I had suddenly left the forbidding coast and come into a smooth little harbor of friendship.

'You ain't never been up to my place,' said the old man.

'Folks don't come now as they used to; no, 't ain't no use to ask folks now. My poor dear she was a great hand to draw young company.'

I remembered that Mrs Todd had once said that this old fisherman had been sore stricken and unconsoled at the death of his wife.

'I should like very much to come,' said I. 'Perhaps you are going to be at home later on?'

Mr Tilley agreed, by a sober nod, and went his way bent-shouldered and with a rolling gait. There was a new patch high on the shoulder of his old waistcoat, which correspond-ed to the renewing of the Miranda's mainsail down the bay, and I wondered if his own fingers, clumsy with much deep-sea fishing, had set it in.

'Was there a good catch to-day?' I asked, stopping a moment. 'I didn't happen to be on the shore when the boats came in.'

'No; all come in pretty light,' answered Mr Tilley. 'Addicks an' Bowden they done the best; Abel an' me we had but a slim fare. We went out 'arly, but not so 'arly as sometimes; looked like a poor mornin'. I got nine haddick, all small, and seven fish; the rest on 'em got more fish than haddick. Well, I don't expect they feel like bitin' every day; we l'arn to humor 'em a little, an' let 'em have their way 'bout it. These plaguey dog-fish kind of worry 'em.' Mr Tilley pronounced the last sen-tence with much sympathy, as if he looked upon himself as a true friend of all the haddock and codfish that lived on the fishing grounds, and so we parted.

Later in the afternoon I went along the beach again until I came to the foot of Mr Tilley's land, and found his rough track across the cobble-stones and rocks to the field edge, where there was a heavy piece of old wreck timber, like a ship's bone, full of treenails. From this a little footpath, nar-row with one man's treading, led up across the small green field that made Mr Tilley's whole estate, except a straggling pasture that tilted on edge up the steep hillside beyond the house and road. I could hear the tinkle-tankle of a cow-bell somewhere among the spruces by which the pasture was being walked over and forested from every side; it was likely to be called the wood lot before long, but the field was unmolested. I could not see a bush or a brier anywhere within its walls, and hardly a stray pebble showed itself. This was most surprising in that country of firm ledges, and scattered stones which all the walls that industry could devise had hardly begun to clear away off the land. In the narrow field I noticed some stout stakes, apparently planted at random in the grass and among the hills of potatoes, but carefully painted yellow and white to match the house, a neat sharp-edged little dwelling, which looked strangely modern for its owner. I should have much sooner believed that the smart young wholesale egg merchant of the Landing was its occupant than Mr Tilley, since a man's house is really but his larger body, and expresses in a way his nature and character.

I went up the field, following the smooth little path to the side door. As for using the front door, that was a matter of great ceremony; the long grass grew close against the high stone step, and a snowberry bush leaned over it, top-heavy with the weight of a morning-glory vine that had managed to take what the fishermen might call a half hitch about the door-knob. Elijah Tilley came to the side door to receive me; he was knitting a blue yarn stocking without looking on, and was warmly dressed for the season in a thick blue flannel shirt with white crockery buttons, a faded waistcoat and trousers heavily patched at the knees. These were not his fishing clothes. There was something delightful in the grasp of his hand, warm and clean, as if it never touched anything but the comfortable woolen yarn, instead of cold sea water and slippery fish.

'What are the painted stakes for, down in the field?' I hastened to ask, and he came out a step or two along the path to see; and looked at the stakes as if his attention were called to them for the first time.

'Folks laughed at me when I first bought this place an' come here to live,' he explained. 'They said 't wa'n't no kind of a field privilege at all; no place to raise anything, all full o' stones. I was aware 't was good land, an' I worked some on it— odd times when I didn't have nothin' else on hand—till I cleared them loose stones all out. You never see a prettier piece than 't is now; now did ye? Well, as for them painted marks, them 's my buoys. I struck on to some heavy rocks that didn't show none, but a plow 'd be liable to ground on 'em, an' so I ketched holt an' buoyed 'em same's you see. They don't trouble me no more 'n if they wa'n't there,'

'You haven't been to sea for nothing,' I said laughing.

'One trade helps another,' said Elijah with an amiable smile. 'Come right in an' set down. Come in an' rest ye,' he exclaimed, and led the way into his comfortable kitchen. The sunshine poured in at the two further windows, and a cat was curled up sound asleep on the table that stood between them. There was a new-looking light oilcloth of a tiled pattern on the floor, and a crockery teapot, large for a household of only one person, stood on the bright stove. I ventured to say that somebody must be a very good housekeeper.

'That's me,' acknowledged the old fisherman with frankness. 'There ain't nobody here but me. I try to keep things looking right, same 's poor dear left 'em. You set down here in this chair, then you can look off an' see the water. None on 'em thought I was goin' to get along alone, no way, but I wa'n't goin' to have my house turned upsi' down an' all changed about; no, not to please nobody. I was the only one knew just how she liked to have things set, poor dear, an' I said I was goin' to make shift, and I have made shift. I'd rather tough it out alone.' And he sighed heavily, as if to sigh were his familiar consolation.

We were both silent for a minute; the old man looked out of the window, as if he had forgotten I was there.

'You must miss her very much?' I said at last.

'I do miss her,' he answered, and sighed again. 'Folks all kep' repeatin' that time would ease me, but I can't find it does. No, I miss her just the same every day.'

'How long is it since she died?' I asked.

'Eight year now, come the first of October. It don't seem near so long. I've got a sister that comes and stops 'long o' me a little spell, spring an' fall, an' odd times if I send after her. I ain't near so good a hand to sew as I be to knit, and she's very quick to set everything to rights. She's a married woman with a family; her son's folks lives at home, an' I can't make no great claim on her time. But it makes me a kind o' good excuse, when I do send, to help her a little; she ain't none too well off. Poor dear always liked her, and we used to contrive our ways together. 'T is full as easy to be alone. I set here an' think it all over, an' think considerable when the weather's bad to go outside. I get so some days it feels as if poor dear might step right back into this kitchen. I keep a watchin' them doors as if she might step in to ary* one. Yes, ma'am, I keep a-lookin' off an' droppin' o' my stitches; that's just how it seems. I can't git over losin' of her no way nor no how. Yes, ma'am, that's just how it seems to me.'

I did not say anything, and he did not look up.

'I git feelin' so sometimes I have to lay everything by an' go out door. She was a sweet pretty creatur' long 's she lived,' the old man added mournfully. 'There's that little rockin' chair o' her 'n, I set an' notice it an' think how strange 't is a creatur' like her should be gone an' that chair be here right in its old place.'

'I wish I had known her; Mrs Todd told me about your wife one day,' I said.

'You'd have liked to come and see her; all the folks did,' said poor Elijah. 'She'd been so pleased to hear everything and see somebody new that took such an int'rest. She had a kind o' gift to make it pleasant for folks. I guess likely Almiry Todd told you she was a pretty woman, especially in her young days; late years, too, she kep' her looks and come to be so pleasant lookin'. There, 't ain't so much matter, I shall be done afore a great while. No; I sha'n't trouble the fish a great sight more.'

The old widower sat with his head bowed over his knitting, as if he were hastily shortening the very thread of time.* The minutes went slowly by. He stopped his work and clasped his hands firmly together. I saw he had forgotten his guest, and I kept the afternoon watch with him. At last he looked up as if but a moment had passed of his continual loneliness.

'Yes, ma'am, I'm one that has seen trouble,' he said, and began to knit again.

The visible tribute of his careful housekeeping, and the

clean bright room which had once enshrined his wife, and now enshrined her memory, was very moving to me; he had no thought for any one else or for any other place. I began to see her myself in her home,—a delicate-looking, faded little woman, who leaned upon his rough strength and affectionate heart, who was always watching for his boat out of this very window, and who always opened the door and welcomed him when he came home.

'I used to laugh at her, poor dear,' said Elijah, as if he read my thought. 'I used to make light of her timid notions. She used to be fearful when I was out in bad weather or baffled about gittin' ashore. She used to say the time seemed long to her, but I've found out all about it now. I used to be dreadful thoughtless when I was a young man and the fish was bitin' well. I'd stay out late some o' them days, an' I expect she'd watch an' watch an' lose heart a-waitin'. My heart alive! what a supper she'd git, an' be right there watchin' from the door, with somethin' over her head if 't was cold, waitin' to hear all about it as I come up the field. Lord, how I think o' all them little things!'

'This was what she called the best room; in this way,' he said presently, laying his knitting on the table, and leading the way across the front entry and unlocking a door, which he threw open with an air of pride. The best room seemed to me a much sadder and more empty place than the kitchen; its conventionalities lacked the simple perfection of the humbler room and failed on the side of poor ambition; it was only when one remembered what patient saving, and what high respect for society in the abstract go to such furnishing that the little parlor was interesting at all. I could imagine the great day of certain purchases, the bewildering shops of the next large town, the aspiring anxious woman, the clumsy sea-tanned man in his best clothes, so eager to be pleased, but at ease only when they were safe back in the sail-boat again, going down the bay with their precious freight, the hoarded money all spent and nothing to think of but tiller and sail. I looked at the unworn carpet, the glass vases on the mantel-piece with their prim bunches of bleached swamp grass and dusty marsh rosemary, and I could read the history of Mrs Tilley's best room from its very beginning.

'You see for yourself what beautiful rugs she could make; now I'm going to show you her best tea things she thought so much of,' said the master of the house, opening the door of a shallow cupboard. 'That's real chiny, all of it on those two shelves,' he told me proudly. 'I bought it all myself, when we was first married, in the port of Bordeaux. There never was one single piece of it broke until—Well, I used to say, long as she lived, there never was a piece broke, but long at the last I noticed she'd look kind o' distressed, an' I thought 't was 'count o' me boastin'. When they asked if they should use it when the folks was here to supper, time o' her funeral, I knew she'd want to have everything nice, and I said "certain." Some o' the women they come runnin' to me an' called me, while they was takin' of the chiny down, an' showed me there was one o' the cups broke an' the pieces wropped in paper and pushed way back here, corner o' the shelf. They didn't want me to go an' think they done it. Poor dear! I had to put right out o' the house when I see that. I knowed in one minute how 't was. We'd got so used to sayin' 't was all there just 's I fetched it home, an' so when she broke that cup somehow or 'nother she couldn't frame no words to come an' tell me. She couldn't think 't would vex me, 't was her own hurt pride. I guess there wa'n't no other secret ever lay between us.'

The French cups with their gay sprigs of pink and blue, the best tumblers, an old flowered bowl and tea caddy, and a japanned waiter or two adorned the shelves. These, with a few daguerreotypes in a little square pile, had the closet to themselves, and I was conscious of much pleasure in seeing them. One is shown over many a house in these days where the interest may be more complex, but not more definite.

'Those were her best things, poor dear,' said Elijah as he locked the door again. 'She told me that last summer before she was taken away that she couldn't think o' anything more she wanted, there was everything in the house, an' all her rooms was furnished pretty. I was goin' over to the Port, an' inquired for errands. I used to ask her to say what she wanted, cost or no cost—she was a very reasonable woman, an' 't was the place where she done all but her extra shopping. It kind o' chilled me up when she spoke so satisfied.'

'You don't go out fishing after Christmas?' I asked, as we came back to the bright kitchen.

'No; I take stiddy to my knitting after January sets in,' said the old seafarer. ''T ain't worth while, fish make off into deeper water an' you can't stand no such perishin' for the sake o' what you get. I leave out a few traps in sheltered coves an' do a little lobsterin' on fair days. The young fellows braves it out, some on 'em; but, for me, I lay in my winter's yarn an' set here where 't is warm, an' knit an' take my comfort. Mother learnt me once when I was a lad; she was a beautiful knitter herself. I was laid up with a bad knee, an' she said 't would take up my time an' help her; we was a large family. They'll buy all the folks can do down here to Addicks' store. They say our Dunnet stockin's is gettin' to be celebrated up to Boston,—good quality o' wool an' even knittin' or somethin'. I've always been called a pretty hand to do nettin', but seines is master cheap to what they used to be when they was all hand worked. I change off to nettin' long towards spring, and I piece up my trawls and lines and get my fishin' stuff to rights. Lobster pots they require attention, but I make 'em up in spring weather when it's warm there in the barn. No; I ain't one o' them that likes to set an' do nothin'.'

'You see the rugs, poor dear did them; she wa'n't very partial to knittin',' old Elijah went on, after he had counted his stitches. 'Our rugs is beginnin' to show wear, but I can't master none o' them womanish tricks. My sister, she tinkers 'em up. She said last time she was here that she guessed they'd last my time.'

'The old ones are always the prettiest,' I said.

'You ain't referrin' to the braided ones now?' answered Mr Tilley. 'You see ours is braided for the most part, an' their good looks is all in the beginnin'. Poor dear used to say they made an easier floor. I go shufflin' round the house same 's if 't was a bo't, and I always used to be stubbin' up the corners o' the hooked kind. Her an' me was always havin' our jokes together same 's a boy an' girl. Outsiders never 'd know nothin' about it to see us. She had nice manners with all, but to me there was nobody so entertainin'. She'd take off anybody's natural talk winter evenin's when we set here alone, so you'd think 't was them a-speakin'. There, there!'

I saw that he had dropped a stitch again, and was snarling the blue yarn round his clumsy fingers. He handled it and threw it off at arm's length as if it were a cod line; and frowned impatiently, but I saw a tear shining on his cheek.

I said that I must be going, it was growing late, and asked if I might come again, and if he would take me out to the fishing grounds some day.

'Yes, come any time you want to,' said my host, ''t ain't so pleasant as when poor dear was here. Oh, I didn't want to lose her an' she didn't want to go, but it had to be. Such things ain't for us to say; there's no yes an' no to it.'

'You find Almiry Todd one o' the best o' women?' said Mr Tilley as we parted. He was standing in the doorway and I had started off down the narrow green field. 'No, there ain't a better hearted woman in the State o' Maine. I've known her from a girl. She's had the best o' mothers. You tell her I'm liable to fetch her up a couple or three nice good mackerel early to-morrow,' he said. 'Now don't let it slip your mind. Poor dear, she always thought a sight o' Almiry, and she used to remind me there was nobody to fish for her; but I don't rec'lect it as I ought to. I see you drop a line yourself very handy now an' then.'

We laughed together like the best of friends, and I spoke again about the fishing grounds, and confessed that I had no fancy for a southerly breeze and a ground swell.

'Nor me neither,' said the old fisherman. 'Nobody likes 'em, say what they may. Poor dear was disobliged by the mere sight of a bo't. Almiry's got the best o' mothers, I expect you know; Mis' Blackett out to Green Island; and we was always plannin' to go out when summer come; but there, I couldn't pick no day's weather that seemed to suit her just right. I never set out to worry her neither, 't wa'n't no kind o' use; she was so pleasant we couldn't have no fret nor trouble. 'T was never "you dear an', you darlin'" afore folks, an' "you divil" behind the door!'

As I looked back from the lower end of the field I saw him still standing, a lonely figure in the doorway. 'Poor dear,' I repeated to myself half aloud; 'I wonder where she is and what she knows of the little world she left. I wonder what she has been doing these eight years!'

I gave the message about the mackerel to Mrs Todd.

'Been visitin' with 'Lijah?' she asked with interest. 'I expect you had kind of a dull session; he ain't the talkin' kind; dwellin' so much long o' fish seems to make 'em lose the gift o' speech.' But when I told her that Mr Tilley had been talking to me that day, she interrupted me quickly.

'Then 't was all about his wife, an' he can't say nothin' too pleasant neither. She was modest with strangers, but there ain't one o' her old friends can ever make up her loss. For me, I don't want to go there no more. There's some folks you miss and some folks you don't, when they're gone, but there ain't hardly a day I don't think o' dear Sarah Tilley. She was always right there; yes, you knew just where to find her like a plain flower. 'Lijah's worthy enough; I do esteem 'Lijah, but he's a ploddin' man.'

XXI

THE BACKWARD VIEW

At last it was the time of late summer, when the house was cool and damp in the morning, and all the light seemed to come through green leaves; but at the first step out of doors the sunshine always laid a warm hand on my shoulder, and the clear, high sky seemed to lift quickly as I looked at it. There was no autumnal mist on the coast, nor any August fog; instead of these, the sea, the sky, all the long shore line and the inland hills, with every bush of bay and every fir-top, gained a deeper color and a sharper clearness. There was something shining in the air, and a kind of lustre on the water and the pasture grass,—a northern look that, except at this moment of the year, one must go far to seek. The sunshine of a northern summer was coming to its lovely end.

The days were few then at Dunnet Landing, and I let each of them slip away unwillingly as a miser spends his coins. I wished to have one of my first weeks back again, with those long hours when nothing happened except the growth of herbs and the course of the sun. Once I had not even known where to go for a walk; now there were many delightful things to be done and done again, as if I were in London. I felt hurried and full of pleasant engagements, and the days flew by like a handful of flowers flung to the sea wind.

At last I had to say good-by to all my Dunnet Landing friends, and my homelike place in the little house, and return to the world in which I feared to find myself a foreigner. There may be restrictions to such a summer's happiness, but the ease that belongs to simplicity is charming enough to make up for whatever a simple life may lack, and the gifts of peace are not for those who live in the thick of battle.

I was to take the small unpunctual steamer that went down the bay in the afternoon, and I sat for a while by my window looking out on the green herb garden, with regret for company. Mrs Todd had hardly spoken all day except in the briefest and most disapproving way; it was as if we were on the edge of a quarrel. It seemed impossible to take my departure with anything like composure. At last I heard a footstep, and looked up to find that Mrs Todd was standing at the door.

'I've seen to everything now,' she told me in an unusually loud and business-like voice. 'Your trunks are on the w'arf by this time. Cap'n Bowden he come and took 'em down himself, an' is going to see that they're safe aboard. Yes, I've seen to all your 'rangements,' she repeated in a gentler tone. 'These things I've left on the kitchen table you'll want to carry by hand; the basket needn't be returned. I guess I shall walk over towards the Port now an' inquire how old Mis' Edward Caplin is.'

I glanced at my friend's face, and saw a look that touched me to the heart. I had been sorry enough before to go away.

'I guess you'll excuse me if I ain't down there to stand round on the w'arf and see you go,' she said, still trying to be gruff. 'Yes, I ought to go over and inquire for Mis' Edward Caplin; it's her third shock, and if mother gets in on Sunday she'll want to know just how the old lady is.' With this last word Mrs Todd turned and left me as if with sudden thought of something she had forgotten, so that I felt sure she was coming back, but presently I heard her go out of the kitchen door and walk down the path toward the gate. I could not part so; I ran after her to say good-by, but she shook her head and waved her hand without looking back when she heard my hurrying steps, and so went away down the street.

When I went in again the little house had suddenly grown lonely, and my room looked empty as it had the day I came. I and all my belongings had died out of it, and I knew how it would seem when Mrs Todd came back and found her lodger gone. So we die before our own eyes; so we see some chapters of our lives come to their natural end.

I found the little packages on the kitchen table. There was a quaint West Indian basket which I knew its owner had valued, and which I had once admired; there was an affecting provision laid beside it for my seafaring supper, with a neatly tied bunch of southernwood and a twig of bay, and a little old leather box which held the coral pin that Nathan Todd brought home to give to poor Joanna.

There was still an hour to wait, and I went up to the hill just above the schoolhouse and sat there thinking of things, and looking off to sea, and watching for the boat to come in sight. I could see Green Island, small and darkly wooded at that distance; below me were the houses of the village with their apple-trees and bits of garden ground. Presently, as I looked at the pastures beyond, I caught a last glimpse of Mrs Todd herself, walking slowly in the footpath that led along, following the shore toward the Port. At such a distance one can feel the large, positive qualities that control a character. Close at hand, Mrs Todd seemed able and warm-hearted and quite absorbed in her bustling industries, but her distant figure looked mateless and appealing, with something about it that was strangely self-possessed and mysterious. Now and then she stooped to pick something,—it might have been her favorite pennyroyal,—and at last I lost sight of her as she slowly crossed an open space on one of the higher points of land, and disappeared again behind a dark clump of juniper and the pointed firs.

As I came away on the little coastwise steamer, there was an old sea running which made the surf leap high on all the rocky shores. I stood on deck, looking back, and watched the busy gulls agree and turn, and sway together down the long slopes of air, then separate hastily and plunge into the waves. The tide was setting in, and plenty of small fish were coming with it, unconscious of the silver flashing of the great birds overhead and the quickness of their fierce beaks. The sea was full of life and spirit, the tops of the waves flew back as if they were winged like the gulls themselves, and like them had the freedom of the wind. Out in the main channel we passed a bent-shouldered old fisherman bound for the evening round among his lobster traps. He was toiling along with short oars, and the dory tossed and sank and tossed again with the steamer's waves. I saw that it was old Elijah Tilley, and though we had so long been strangers we had come to be warm friends, and I wished that he had waited for one of his mates, it was such hard work to row along shore through rough seas and tend the traps alone. As we passed I waved my hand and tried to call to him, and he looked up and answered my farewells by a solemn nod. The little town, with the tall masts of its disabled schooners in the inner bay, stood high above the flat sea for a few minutes, then it sank back into the uniformity of the coast, and became indistinguishable from the other towns that looked as if they were crumbled on the furzy-green stoniness of the shore.

The small outer islands of the bay were covered among the ledges with turf that looked as fresh as the early grass; there had been some days of rain the week before, and the darker green of the sweet-fern was scattered on all the pasture heights. It looked like the beginning of summer ashore, though the sheep, round and warm in their winter wool, betrayed the season of the year as they went feeding along the slopes in the low afternoon sunshine. Presently the wind began to blow, and we struck out seaward to double the long sheltering headland of the cape, and when I looked back again, the islands and the headland had run together and Dunnet Landing and all its coasts were lost to sight.

DUNNET LANDING
STORIES

THE QUEEN'S TWIN

I

THE coast of Maine was in former years brought so near to foreign shores by its busy fleet of ships that among the older men and women one still finds a surprising proportion of travelers. Each seaward-stretching headland with its high-set houses, each island of a single farm, has sent its spies to view many a Land of Eshcol;* one may see plain, contented old faces at the windows, whose eyes have looked at far-away ports and known the splendors of the Eastern world. They shame the easy voyager of the North Atlantic and the Mediterranean; they have rounded the Cape of Good Hope and braved the angry seas of Cape Horn in small wooden ships; they have brought up their hardy boys and girls on narrow decks; they were among the last of the Northmen's children to go adventuring to unknown shores. More than this one cannot give to a young State for its enlightenment; the sea captains and the captains' wives of Maine knew something of the wide world, and never mistook their native parishes for the whole instead of a part thereof; they knew not only Thomaston and Castine and Portland, but London and Bristol and Bordeaux, and the strange-mannered harbors of the China Sea.

One September day, when I was nearly at the end of a summer spent in a village called Dunnet Landing, on the Maine coast, my friend Mrs Todd, in whose house I lived, came home from a long, solitary stroll in the wild pastures, with an eager look as if she were just starting on a hopeful quest instead of returning. She brought a little basket with blackberries enough for supper, and held it towards me so that I could see that there were also some late and surprising raspberries sprinkled on top, but she made no comment upon her wayfaring. I could tell plainly that she had something very important to say.

'You haven't brought home a leaf of anything,' I ventured

to this practiced herb-gatherer. 'You were saying yesterday that the witch hazel might be in bloom.'

'I dare say, dear,' she answered in a lofty manner; 'I ain't goin' to say it wasn't; I ain't much concerned either way 'bout the facts o' witch hazel. Truth is, I've been off visitin'; there's an old Indian footpath leadin' over towards the Back Shore through the great heron swamp that anybody can't travel over all summer. You have to seize your time some day just now, while the low ground's summer-dried as it is to-day, and before the fall rains set in. I never thought of it till I was out o' sight o' home, and I says to myself, "To-day's the day, certain!" and stepped along smart as I could. Yes, I've been visitin'. I did get into one spot that was wet underfoot before I noticed; you wait till I get me a pair o' dry woolen stockings, in case of cold, and I'll come an' tell ye.'

Mrs Todd disappeared. I could see that something had deeply interested her. She might have fallen in with either the sea-serpent or the lost tribes of Israel,* such was her air of mystery and satisfaction. She had been away since just before mid-morning, and as I sat waiting by my window I saw the last red glow of autumn sunshine flare along the gray rocks of the shore and leave them cold again, and touch the far sails of some coastwise schooners so that they stood like golden houses on the sea.

I was left to wonder longer than I liked. Mrs Todd was making an evening fire and putting things in train for supper; presently she returned, still looking warm and cheerful after her long walk.

'There's a beautiful view from a hill over where I've been,' she told me; 'yes, there's a beautiful prospect of land and sea. You wouldn't discern the hill from any distance, but 't is the pretty situation of it that counts. I sat there a long spell, and I did wish for you. No, I didn't know a word about goin' when I set out this morning' (as if I had openly reproached her!); 'I only felt one o' them travelin' fits comin' on, an' I ketched up my little basket; I didn't know but I might turn and come back time for dinner. I thought it wise to set out your luncheon for you in case I didn't. Hope you had all you wanted; yes, I hope you had enough.'

'Oh, yes, indeed,' said I. My landlady was always peculiarly bountiful in her supplies when she left me to fare for myself, as if she made a sort of peace-offering or affectionate apology.

'You know that hill with the old house right on top, over beyond the heron swamp? You'll excuse me for explainin',' Mrs Todd began, 'but you ain't so apt to strike inland as you be to go right along shore. You know that hill; there's a path leadin' right over to it that you have to look sharp to find nowadays; it belonged to the up-country Indians when they had to make a carry to the landing here to get to the out' islands. I've heard the old folks say that there used to be a place across a ledge where they'd worn a deep track with their moccasin feet, but I never could find it. 'T is so overgrown in some places that you keep losin' the path in the bushes and findin' it as you can; but it runs pretty straight considerin' the lay o' the land, and I keep my eye on the sun and the moss that grows one side o' the tree trunks. Some brook's been choked up and the swamp's bigger than it used to be. Yes; I did get in deep enough, one place!'

I showed the solicitude that I felt. Mrs Todd was no longer young, and in spite of her strong, great frame and spirited behavior, I knew that certain ills were apt to seize upon her, and would end some day by leaving her lame and ailing.

'Don't you go to worryin' about me,' she insisted, 'settin' still's the only way the Evil One'll* ever get the upper hand o' me. Keep me movin' enough, an' I'm twenty year old summer an' winter both. I don't know why 't is, but I've never happened to mention the one I've been to see. I don't know why I never happened to speak the name of Abby Martin, for I often give her a thought, but 't is a dreadful out-o'-the-way place where she lives, and I haven't seen her myself for three or four years. She's a real good interesting woman, and we're well acquainted; she's nigher mother's age than mine, but she's very young feeling. She made me a nice cup o' tea, and I don't know but I should have stopped all night if I could have got word to you not to worry.'

Then there was a serious silence before Mrs Todd spoke again to make a formal announcement.

'She is the Queen's Twin,' and Mrs Todd looked steadily to see how I might bear the great surprise.

'The Queen's Twin?' I repeated.

'Yes, she's come to feel a real interest in the Queen, and anybody can see how natural 't is. They were born the very same day,* and you would be astonished to see what a number o' other things have corresponded. She was speaking o' some o' the facts to me to-day, an' you 'd think she'd never done nothing but read history. I see how earnest she was about it as I never did before. I've often and often heard her allude to the facts, but now she's got to be old and the hurry's over with her work, she's come to live a good deal in her thoughts, as folks often do, and I tell you 't is a sight o' company for her. If you want to hear about Queen Victoria, why Mis' Abby Martin'll tell you everything. And the prospect from that hill I spoke of is as beautiful as anything in this world; 't is worth while your goin' over to see her just for that.'

'When can you go again?' I demanded eagerly.

'I should say to-morrow,' answered Mrs Todd; 'yes, I should say to-morrow; but I expect 't would be better to take one day to rest, in between. I considered that question as I was comin' home, but I hurried so that there wa'n't much time to think. It's a dreadful long way to go with a horse; you have to go 'most as far as the old Bowden place an' turn off to the left, a master long, rough road, and then you have to turn right round as soon as you get there if you mean to get home before nine o'clock at night. But to strike across country from here, there's plenty o' time in the shortest day, and you can have a good hour or two's visit beside; 't ain't but a very few miles, and it's pretty all the way along. There used to be a few good families over there, but they've died and scattered, so now she's far from neighbors. There, she really cried, she was so glad to see anybody comin'. You'll be amused to hear her talk about the Queen, but I thought twice or three times as I set there 't was about all the company she 'd got.'

'Could we go day after to-morrow?' I asked eagerly.

' 'T would suit me exactly,' said Mrs Todd.

II

ONE can never be so certain of good New England weather as in the days when a long easterly storm has blown away the warm late-summer mists, and cooled the air so that however bright the sunshine is by day, the nights come nearer and nearer to frostiness. There was a cold freshness in the morning air when Mrs Todd and I locked the house-door behind us; we took the key of the fields into our own hands that day, and put out across country as one puts out to sea. When we reached the top of the ridge behind the town it seemed as if we had anxiously passed the harbor bar and were comfortably in open sea at last.

'There, now!' proclaimed Mrs Todd, taking a long breath, 'now I do feel safe. It's just the weather that's liable to bring somebody to spend the day; I've had a feeling of Mis' Elder Caplin from North Point bein' close upon me ever since I waked up this mornin', an' I didn't want to be hampered with our present plans. She's a great hand to visit; she'll be spendin' the day somewhere from now till Thanksgivin', but there's plenty o' places at the Landin' where she goes, an' if I ain't there she'll just select another. I thought mother might be in, too, 't is so pleasant; but I run up the road to look off this mornin' before you was awake, and there was no sign o' the boat. If they hadn't started by that time they wouldn't start, just as the tide is now; besides, I see a lot o' mackerel-men headin' Green Island way, and they'll detain William. No, we're safe now, an' if mother should be comin' in tomorrow we'll have all this to tell her. She an' Mis' Abby Martin's very old friends.'

We were walking down the long pasture slopes towards the dark woods and thickets of the low ground. They stretched away northward like an unbroken wilderness; the early mists still dulled much of the color and made the uplands beyond look like a very far-off country.

'It ain't so far as it looks from here,' said my companion reassuringly, 'but we've got no time to spare either,' and she hurried on, leading the way with a fine sort of spirit in her step; and presently we struck into the old Indian footpath, which could be plainly seen across the long-unploughed turf of the pastures, and followed it among the thick, low-growing spruces. There the ground was smooth and brown under foot, and the thin-stemmed trees held a dark and shadowy roof overhead. We walked a long way without speaking; sometimes we had to push aside the branches, and sometimes we walked in a broad aisle where the trees were larger. It was a solitary wood, birdless and beastless; there was not even a rabbit to be seen, or a crow high in air to break the silence.

'I don't believe the Queen ever saw such a lonesome trail as this,' said Mrs Todd, as if she followed the thoughts that were in my mind. Our visit to Mrs Abby Martin seemed in some strange way to concern the high affairs of royalty. I had just been thinking of English landscapes, and of the solemn hills of Scotland with their lonely cottages and stone-walled sheep-folds, and the wandering flocks on high cloudy pastures. I had often been struck by the quick interest and familiar allusion to certain members of the royal house which one found in distant neighborhoods of New England; whether some old instincts of personal loyalty have survived all changes of time and national vicissitudes, or whether it is only that the Queen's own character and disposition have won friends for her so far away, it is impossible to tell. But to hear of a twin sister was the most surprising proof of intimacy of all, and I must confess that there was something remarkably exciting to the imagination in my morning walk. To think of being presented at Court in the usual way was for the moment quite commonplace.

III

MRS TODD was swinging her basket to and fro like a school-girl as she walked, and at this moment it slipped from her hand and rolled lightly along the ground as if there were nothing in it. I picked it up and gave it to her, whereupon she lifted the cover and looked in with anxiety.

''T is only a few little things, but I don't want to lose 'em,' she explained humbly. ''T was lucky you took the other basket if I was goin' to roll it round. Mis' Abby Martin complained o' lacking some pretty pink silk to finish one o' her little frames, an' I thought I'd carry her some, and I had a bunch o' gold thread that had been in a box o' mine this twenty year. I never was one to do much fancy work, but we're all liable to be swept away by fashion. And then there's a small packet o' very choice herbs that I gave a good deal of attention to; they'll smarten her up and give her the best of appetites, come spring. She was tellin' me that spring weather is very wiltin' an' tryin' to her, and she was beginnin' to dread it already. Mother 's just the same way; if I could prevail on mother to take some o' these remedies in good season 't would make a world o' difference, but she gets all down hill before I have a chance to hear of it, and then William comes in to tell me, sighin' and bewailin', how feeble mother is. "Why can't you remember 'bout them good herbs that I never let her be without?" I say to him—he does provoke me so; and then off he goes, sulky enough, down to his boat. Next thing I know, she comes in to go to meetin', wantin' to speak to everybody and feelin' like a girl. Mis' Martin's case is very much the same; but she's nobody to watch her. William's kind o' slow-moulded; but there, any William's better than none when you get to be Mis' Martin's age.'

'Hadn't she any children?' I asked.

'Quite a number,' replied Mrs Todd grandly, 'but some are gone and the rest are married and settled. She never was a great hand to go about visitin'. I don't know but Mis' Martin might be called a little peculiar. Even her own folks has to make company of her; she never slips in and lives right along with the rest as if 't was at home, even in her own children's houses. I heard one o' her sons' wives say once she'd much rather have the Queen to spend the day if she could choose between the two, but I never thought Abby was so difficult as that. I used to love to have her come; she may have been sort o' ceremonious, but very pleasant and sprightly if you had sense enough to treat her her own way. I always think she'd know just how to live with great folks, and feel easier 'long of them an' their ways. Her son's wife's a great driver with farm-work, boards a great tableful o' men in hayin' time, an' feels right in her element. I don't say but she's a good woman an' smart, but sort o' rough. Anybody that's gentle-mannered an' precise like Mis' Martin would be a sort o' restraint.

'There's all sorts o' folks in the country, same 's there is in the city,' concluded Mrs Todd gravely, and I as gravely agreed. The thick woods were behind us now, and the sun was shining clear overhead, the morning mists were gone, and a faint blue haze softened the distance; as we climbed the hill where we were to see the view, it seemed like a summer day. There was an old house on the height, facing southward,—a mere forsaken shell of an old house, with empty windows that looked like blind eyes. The frost-bitten grass grew close about it like brown fur, and there was a single crooked bough of lilac holding its green leaves close by the door.

'We'll just have a good piece of bread-an'-butter now,' said the commander of the expedition, 'and then we'll hang up the basket on some peg inside the house out o' the way o' the sheep, and have a han'some entertainment as we're comin' back. She'll be all through her little dinner when we get there, Mis' Martin will; but she'll want to make us some tea, an' we must have our visit an' be startin' back pretty soon after two. I don't want to cross all that low ground again after it's begun to grow chilly. An' it looks to me as if the clouds might begin to gather late in the afternoon.'

Before us lay a splendid world of sea and shore. The autumn colors already brightened the landscape; and here and there at the edge of a dark tract of pointed firs stood a row of bright swamp-maples like scarlet flowers. The blue sea and the great tide inlets were untroubled by the lightest winds.

'Poor land, this is!' sighed Mrs Todd as we sat down to rest on the worn doorstep. 'I've known three good hard-workin' families that come here full o' hope an' pride and tried to make something o' this farm, but it beat 'em all. There's one small field that's excellent for potatoes if you let half of it rest every year; but the land's always hungry. Now, you see them little peakéd-topped spruces an' fir balsams comin' up over the hill all green an' hearty; they've got it all their own way! Seems sometimes as if wild Natur' got jealous over a certain spot, and wanted to do just as she'd a mind to. You'll see here;

she'll do her own ploughin' an' harrowin' with frost an' wet, an' plant just what she wants and wait for her own crops. Man can't do nothin' with it, try as he may. I tell you those little trees means business!'

I looked down the slope, and felt as if we ourselves were likely to be surrounded and overcome if we lingered too long. There was a vigor of growth, a persistence and savagery about the sturdy little trees that put weak human nature at complete defiance. One felt a sudden pity for the men and women who had been worsted after a long fight in that lonely place; one felt a sudden fear of the unconquerable, immediate forces of Nature, as in the irresistible moment of a thunderstorm.

'I can recollect the time when folks were shy o' these woods we just come through,' said Mrs Todd seriously. 'The menfolks themselves never'd venture into 'em alone; if their cattle got strayed they'd collect whoever they could get, and start off all together. They said a person was liable to get bewildered in there alone, and in old times folks had been lost. I expect there was considerable fear left over from the old Indian times, and the poor days o' witchcraft; anyway, I've seen bold men act kind o' timid. Some women o' the Asa Bowden family went out one afternoon berryin' when I was a girl, and got lost and was out all night; they found 'em middle o' the mornin' next day, not half a mile from home, scared most to death, an' sayin' they'd heard wolves and other beasts sufficient for a caravan. Poor creatur's! they'd strayed at last into a kind of low place amongst some alders, an' one of 'em was so overset she never got over it, an' went off in a sort o' slow decline. 'T was like them victims that drowns in a foot o' water; but their minds did suffer dreadful. Some folks is born afraid of the woods and all wild places, but I must say they've always been like home to me.'

I glanced at the resolute, confident face of my companion. Life was very strong in her, as if some force of Nature were personified in this simple-hearted woman and gave her cousinship to the ancient deities. She might have walked the primeval fields of Sicily; her strong gingham skirts might at that very moment bend the slender stalks of asphodel and be fragrant with trodden thyme, instead of the brown wind-brushed grass of New England and frost-bitten goldenrod. She was a great soul, was Mrs Todd, and I her humble follower, as we went our way to visit the Queen's Twin, leaving the bright view of the sea behind us, and descending to a lower country-side through the dry pastures and fields.

The farms all wore a look of gathering age, though the settlement was, after all, so young. The fences were already fragile, and it seemed as if the first impulse of agriculture had soon spent itself without hope of renewal. The better houses were always those that had some hold upon the riches of the sea; a house that could not harbor a fishing-boat in some neighboring inlet was far from being sure of every-day comforts. The land alone was not enough to live upon in that stony region; it belonged by right to the forest, and to the forest it fast returned. From the top of the hill where we had been sitting we had seen prosperity in the dim distance, where the land was good and the sun shone upon fat barns, and where warm-looking houses with three or four chimneys apiece stood high on their solid ridge above the bay.

As we drew nearer to Mrs Martin's it was sad to see what poor bushy fields, what thin and empty dwelling-places had been left by those who had chosen this disappointing part of the northern country for their home. We crossed the last field and came into a narrow rain-washed road, and Mrs Todd looked eager and expectant and said that we were almost at our journey's end. 'I do hope Mis' Martin'll ask you into her best room where she keeps all the Queen's pictures. Yes, I think likely she will ask you; but 't ain't everybody she deems worthy to visit 'em, I can tell you!' said Mrs Todd warningly. 'She's been collectin' 'em an' cuttin' 'em out o' newspapers an' magazines time out o' mind, and if she heard of anybody sailin' for an English port she 'd contrive to get a little money to 'em and ask to have the last likeness there was. She 's most covered her best-room wall now; she keeps that room shut up sacred as a meetin'-house! "I won't say but I have my favorites amongst 'em," she told me t' other day, "but they're all beautiful to me as they can be!" And she 's made some kind o' pretty little frames for 'em all—you know there's always a new fashion o' frames comin' round; first 't was shell-work, and then 't was pine-cones, and bead-work's had its day, and now she's much concerned with perforated cardboard worked with silk. I tell you that best room's a sight to see! But you mustn't look for anything elegant,' continued Mrs Todd, after a moment's reflection. 'Mis' Martin's always been in very poor, strugglin' circumstances. She had ambition for her children, though they took right after their father an' had little for themselves; she wa'n't over an' above well married, however kind she may see fit to speak. She's been patient an' hard-workin' all her life, and always high above makin' mean complaints of other folks. I expect all this business about the Queen has buoyed her over many a shoal place in life. Yes, you might say that Abby'd been a slave, but there ain't any slave but has some freedom.'

IV

PRESENTLY I saw a low gray house standing on a grassy bank close to the road. The door was at the side, facing us, and a tangle of snowberry bushes and cinnamon roses grew to the level of the window-sills. On the doorstep stood a bent-shouldered, little old woman; there was an air of welcome and of unmistakable dignity about her.

'She sees us coming,' exclaimed Mrs Todd in an excited whisper. 'There, I told her I might be over this way again if the weather held good, and if I came I'd bring you. She said right off she'd take great pleasure in havin' a visit from you: I was surprised, she's usually so retirin'.'

Even this reassurance did not quell a faint apprehension on our part; there was something distinctly formal in the occasion, and one felt that consciousness of inadequacy which is never easy for the humblest pride to bear. On the way I had torn my dress in an unexpected encounter with a little thorn-bush, and I could now imagine how it felt to be going to Court and forgetting one's feathers or her Court train.

The Queen's Twin was oblivious of such trifles; she stood waiting with a calm look until we came near enough to take her kind hand. She was a beautiful old woman, with clear eyes and a lovely quietness and genuineness of manner; there was not a trace of anything pretentious about her, or high-flown, as Mrs Todd would say comprehensively. Beauty in age is rare enough in women who have spent their lives in the hard work of a farmhouse; but autumn-like and withered as this woman may have looked, her features had kept, or rather gained, a great refinement. She led us into her old kitchen and gave us seats, and took one of the little straight-backed chairs herself and sat a short distance away, as if she were giving audience to an ambassador. It seemed as if we should all be standing; you could not help feeling that the habits of her life were more ceremonious, but that for the moment she assumed the simplicities of the occasion.

Mrs Todd was always Mrs Todd, too great and self-possessed a soul for any occasion to ruffle. I admired her calmness, and presently the slow current of neighborhood talk carried one easily along; we spoke of the weather and the small adventures of the way, and then, as if I were after all not a stranger, our hostess turned almost affectionately to speak to me.

'The weather will be growing dark in London now. I expect that you've been in London, dear?' she said.

'Oh, yes,' I answered. 'Only last year.'

'It is a great many years since I was there, along in the forties,' said Mrs Martin. ''T was the only voyage I ever made; most of my neighbors have been great travelers. My brother was master of a vessel, and his wife usually sailed with him; but that year she had a young child more frail than the others, and she dreaded the care of it at sea. It happened that my brother got a chance for my husband to go as supercargo, being a good accountant, and came one day to urge him to take it; he was very ill-disposed to the sea, but he had met with losses, and I saw my own opportunity and persuaded them both to let me go too. In those days they didn't object to a woman's being aboard to wash and mend, the voyages were sometimes very long. And that was the way I come to see the Queen.'

Mrs Martin was looking straight in my eyes to see if I showed any genuine interest in the most interesting person in the world.

'Oh, I am very glad you saw the Queen,' I hastened to say. 'Mrs Todd has told me that you and she were born the very same day.'

'We were indeed, dear!' said Mrs Martin, and she leaned back comfortably and smiled as she had not smiled before. Mrs Todd gave a satisfied nod and glance, as if to say that things were going on as well as possible in this anxious moment.

'Yes,' said Mrs Martin again, drawing her chair a little nearer, "t was a very remarkable thing; we were born the same day, and at exactly the same hour, after you allowed for all the difference in time. My father figured it out sea-fashion. Her Royal Majesty and I opened our eyes upon this world together; say what you may, 't is a bond between us.'

Mrs Todd assented with an air of triumph, and untied her hat-strings and threw them back over her shoulders with a gallant air.

'And I married a man by the name of Albert, just the same as she did, and all by chance, for I didn't get the news that she had an Albert too till a fortnight afterward; news was slower coming then than it is now. My first baby was a girl, and I called her Victoria after my mate; but the next one was a boy, and my husband wanted the right to name him, and took his own name and his brother Edward's, and pretty soon I saw in the paper that the little Prince o' Wales had been christened just the same. After that I made excuse to wait till I knew what she'd named her children. I didn't want to break the chain, so I had an Alfred, and my darling Alice that I lost long before she lost hers, and there I stopped. If I'd only had a dear daughter to stay at home with me, same 's her youngest one, I should have been so thankful! But if only one of us could have a little Beatrice, I'm glad 't was the Queen; we've both seen trouble, but she's had the most care.'

I asked Mrs Martin if she lived alone all the year, and was told that she did except for a visit now and then from one of her grandchildren, 'the only one that really likes to come an' stay quiet 'long o' grandma. She always says quick as she's through her schoolin' she's goin' to live with me all the time, but she's very pretty an' has taking ways,' said Mrs Martin, looking both proud and wistful, 'so I can tell nothing at all about it! Yes, I've been alone most o' the time since my Albert was taken away, and that's a great many years; he had a long time o' failing and sickness first.' (Mrs Todd's foot gave an impatient scuff on the floor.) 'An' I've always lived right here. I ain't like the Queen's Majesty, for this is the only palace I've got,' said the dear old thing, smiling again. 'I'm glad of it too, I don't like changing about, an' our stations in life are set very different. I don't require what the Queen does, but some-times I've thought 't was left to me to do the plain things she don't have time for. I expect she's a beautiful housekeeper, nobody couldn't have done better in her high place, and she's been as good a mother as she's been a queen.'

'I guess she has, Abby,' agreed Mrs Todd instantly. 'How was it you happened to get such a good look at her? I meant to ask you again when I was here t' other day.'

'Our ship was layin' in the Thames, right there above Wap-ping. We was dischargin' cargo, and under orders to clear as quick as we could for Bordeaux to take on an excellent freight o' French goods,' explained Mrs Martin eagerly. 'I heard that the Queen was goin' to a great review of her army, and would drive out o' her Buckin'ham Palace about ten o'clock in the mornin', and I run aft to Albert, my husband, and brother Horace where they was standin' together by the hatchway, and told 'em they must one of 'em take me. They laughed, I was in such a hurry, and said they couldn't go; and I found they meant it and got sort of impatient when I began to talk, and I was 'most broken-hearted; 't was all the reason I had for makin' that hard voyage. Albert couldn't help often re-proachin' me, for he did so resent the sea, an' I'd known how 't would be before we sailed; but I'd minded nothing all the way till then, and I just crep' back to my cabin an' begun to cry. They was disappointed about their ship's cook, an' I'd cooked for fo'c's'le an' cabin myself all the way over; 't was dreadful hard work, specially in rough weather; we'd had head winds an' a six weeks' voyage. They'd acted sort of ashamed o' me when I pled so to go ashore, an' that hurt my feelin's most of all. But Albert come below pretty soon; I'd never given way so in my life, an' he begun to act frightened, and treated me gentle just as he did when we was goin' to be married, an' when I got over sobbin' he went on deck and saw Horace an' talked it over what they could do; they really had their duty to the vessel, and couldn't be spared that day. Horace was real good when he understood everything, and he come an' told me I'd more than worked my passage an' was goin' to do just as I liked now we was in port. He'd engaged a cook, too, that was comin' aboard that mornin', and he was goin' to send the ship's carpenter with me—a nice fellow from up Thomaston way; he'd gone to put on his ashore clothes as quick 's he could. So then I got ready, and we started off in the small boat and rowed up river. I was afraid we were too late, but the tide was setting up very strong, and we landed an' left the boat to a keeper, and I run all the way up those great streets and across a park. 'T was a great day, with sights o' folks everywhere, but 't was just as if they was nothin' but wax images to me. I kep' askin' my way an' runnin' on, with the carpenter comin' after as best he could, and just as I worked to the front o' the crowd by the palace, the gates was flung open and out she came; all prancin' horses and shinin' gold, and in a beautiful carriage there she sat; 't was a moment o' heaven to me. I saw her plain, and she looked right at me so pleasant and happy, just as if she knew there was somethin' different between us from other folks.'

There was a moment when the Queen's Twin could not go on and neither of her listeners could ask a question.

'Prince Albert was sitting right beside her in the carriage,' she continued. 'Oh, he was a beautiful man! Yes, dear, I saw 'em both together just as I see you now, and then she was gone out o' sight in another minute, and the common crowd was all spread over the place pushin' an' cheerin'. 'T was some kind o' holiday, an' the carpenter and I got separated, an' then I found him again after I didn't think I should, an' he was all for makin' a day of it, and goin' to show me all the sights; he'd been in London before, but I didn't want nothin' else, an' we went back through the streets down to the water-side an' took the boat. I remember I mended an old coat o' my Albert's as good as I could, sittin' on the quarter-deck in the sun all that afternoon, and 't was all as if I was livin' in a lovely dream. I don't know how to explain it, but there hasn't been no friend I've felt so near to me ever since.'

One could not say much—only listen. Mrs Todd put in a discerning question now and then, and Mrs Martin's eyes shone brighter and brighter as she talked. What a lovely gift of imagination and true affection was in this fond old heart! I looked about the plain New England kitchen, with its wood-smoked walls and homely braided rugs on the worn floor, and all its simple furnishings. The loud-ticking clock seemed to encourage us to speak; at the other side of the room was an early newspaper portrait of Her Majesty the Queen of Great Britain and Ireland. On a shelf below were some flowers in a little glass dish, as if they were put before a shrine.

'If I could have had more to read, I should have known 'most everything about her,' said Mrs Martin wistfully. 'I've made the most of what I did have, and thought it over and over till it came clear. I sometimes seem to have her all my own, as if we'd lived right together. I've often walked out into the woods alone and told her what my troubles was, and it always seemed as if she told me 't was all right, an' we must have patience. I've got her beautiful book about the Highlands; 't was dear Mis' Todd here that found out about her printing it and got a copy for me, and it's been a treasure to my heart, just as if 't was written right to me. I always read it Sundays now, for my Sunday treat. Before that I used to have to imagine a good deal, but when I come to read her book, I knew what I expected was all true. We do think alike about so many things,' said the Queen's Twin with affectionate certainty. 'You see, there is something between us, being born just at the same time; 't is what they call a birthright. She's had great tasks put upon her, being the Queen, an' mine has been the humble lot; but she's done the best she could, nobody can say to the contrary, and there's something between us; she's been the great lesson I've had to live by. She's been everything to me. An' when she had her Jubilee, oh, how my heart was with her!'

'There, 't wouldn't play the part in her life it has in mine,' said Mrs Martin generously, in answer to something one of her listeners had said. 'Sometimes I think, now she's older, she might like to know about us. When I think how few old friends anybody has left at our age, I suppose it may be just the same with her as it is with me; perhaps she would like to know how we came into life together. But I've had a great advantage in seeing her, an' I can always fancy her goin' on, while she don't know nothin' yet about me, except she may feel my love stayin' her heart sometimes an' not know just where it comes from. An' I dream about our being together out in some pretty fields, young as ever we was, and holdin' hands as we walk along. I'd like to know if she ever has that dream too. I used to have days when I made believe she did know, an' was comin' to see me,' confessed the speaker shyly, with a little flush on her cheeks; 'and I'd plan what I could have nice for supper, and I wasn't goin' to let anybody know she was here havin' a good rest, except I'd wish you, Almira Todd, or dear Mis' Blackett would happen in, for you'd know just how to talk with her. You see, she likes to be up in Scotland, right out in the wild country, better than she does anywhere else.'

'I'd really love to take her out to see mother at Green Island,' said Mrs Todd with a sudden impulse.

'Oh, yes! I should love to have you,' exclaimed Mrs Martin, and then she began to speak in a lower tone. 'One day I got thinkin' so about my dear Queen,' she said, 'an' livin' so in my thoughts, that I went to work an' got all ready for her, just as if she was really comin'. I never told this to a livin' soul before, but I feel you'll understand. I put my best fine sheets and blankets I spun an' wove myself on the bed, and I picked some pretty flowers and put 'em all round the house, an' I worked as hard an' happy as I could all day, and had as nice a supper ready as I could get, sort of telling myself a story all the time. She was comin' an' I was goin' to see her again, an' I kep' it up until nightfall; an' when I see the dark an' it come to me I was all alone, the dream left me, an' I sat down on the doorstep an' felt all foolish an' tired. An', if you'll believe it, I heard steps comin', an' an old cousin o' mine come wander-in' along, one I was apt to be shy of. She wasn't all there, as folks used to say, but harmless enough and a kind of poor old talking body. And I went right to meet her when I first heard her call, 'stead o' hidin' as I sometimes did, an' she come in dreadful willin', an' we sat down to supper together; 't was a supper I should have had no heart to eat alone.'

'I don't believe she ever had such a splendid time in her life as she did then. I heard her tell all about it afterwards,' exclaimed Mrs Todd compassionately. 'There, now I hear all this it seems just as if the Queen might have known and couldn't come herself, so she sent that poor old creatur' that was always in need!'

Mrs Martin looked timidly at Mrs Todd and then at me. '' T was childish o' me to go an' get supper,' she confessed.

'I guess you wa'n't the first one to do that,' said Mrs Todd. 'No, I guess you wa'n't the first one who's got supper that way, Abby,' and then for a moment she could say no more.

Mrs Todd and Mrs Martin had moved their chairs a little so that they faced each other, and I, at one side, could see them both.

'No, you never told me o' that before, Abby,' said Mrs Todd gently. 'Don't it show that for folks that have any fancy in 'em, such beautiful dreams is the real part o' life? But to most folks the common things that happens outside 'em is all in all.'

Mrs Martin did not appear to understand at first, strange to say, when the secret of her heart was put into words; then a glow of pleasure and comprehension shone upon her face. 'Why, I believe you're right, Almira!' she said, and turned to me.

'Wouldn't you like to look at my pictures of the Queen?' she asked, and we rose and went into the best room.

V

THE mid-day visit seemed very short; September hours are brief to match the shortening days. The great subject was dismissed for a while after our visit to the Queen's pictures, and my companions spoke much of lesser persons until we drank the cup of tea which Mrs Todd had foreseen. I happily remembered that the Queen herself is said to like a proper cup of tea, and this at once seemed to make her Majesty kindly join so remote and reverent a company. Mrs Martin's thin cheeks took on a pretty color like a girl's. 'Somehow I always have thought of her when I made it extra good,' she said. 'I've got a real china cup that belonged to my grandmother, and I believe I shall call it hers now.'

'Why don't you?' responded Mrs Todd warmly, with a delightful smile.

Later they spoke of a promised visit which was to be made in the Indian summer to the Landing and Green Island, but I observed that Mrs Todd presented the little parcel of dried herbs, with full directions, for a cure-all in the spring, as if there were no real chance of their meeting again first. As we looked back from the turn of the road the Queen's Twin was still standing on the doorstep watching us away, and Mrs Todd stopped, and stood still for a moment before she waved her hand again.

'There's one thing certain, dear,' she said to me with great discernment; 'It ain't as if we left her all alone!'

Then we set out upon our long way home over the hill, where we lingered in the afternoon sunshine, and through the dark woods across the heron-swamp.

A DUNNET SHEPHERDESS

I

EARLY one morning at Dunnet Landing, as if it were still night, I waked, suddenly startled by a spirited conversation beneath my window. It was not one of Mrs Todd's morning soliloquies; she was not addressing her plants and flowers in words of either praise or blame. Her voice was declamatory though perfectly good-humored, while the second voice, a man's, was of lower pitch and somewhat deprecating.

The sun was just above the sea, and struck straight across my room through a crack in the blind. It was a strange hour for the arrival of a guest, and still too soon for the general run of business, even in that tiny eastern haven where daybreak fisheries and early tides must often rule the day.

The man's voice suddenly declared itself to my sleepy ears. It was Mr William Blackett's.

'Why, sister Almiry,' he protested gently, 'I don't need none o' your nostrums!'

'Pick me a small han'ful,' she commanded. 'No, no, a *small* han'ful, I said,—o' them large pennyr'yal sprigs! I go to all the trouble an' cossetin' of 'em just so as to have you ready to meet such occasions, an' last year, you may remember, you never stopped here at all the day you went up country. An' the frost come at last an' blacked it. I never saw any herb that so objected to gardin ground; might as well try to flourish may-flowers in a common front yard. There, you can come in now, an' set and eat what breakfast you've got patience for. I've found everything I want, an' I'll mash 'em up an' be all ready to put 'em on.'

I heard such a pleading note of appeal as the speakers went round the corner of the house, and my curiosity was so demanding, that I dressed in haste, and joined my friends a little later, with two unnoticed excuses of the beauty of the morning, and the early mail boat. William's breakfast had been slighted; he had taken his cup of tea and merely pushed back the rest on the kitchen table. He was now sitting in a helpless condition by the side window, with one of his sister's purple calico aprons pinned close about his neck. Poor William was meekly submitting to being smeared, as to his countenance, with a most pungent and unattractive lotion of pennyroyal and other green herbs which had been hastily pounded and mixed with cream in the little white stone mortar.

I had to cast two or three straightforward looks at William to reassure myself that he really looked happy and expectant in spite of his melancholy circumstances, and was not being overtaken by retribution. The brother and sister seemed to be on delightful terms with each other for once, and there was something of cheerful anticipation in their morning talk. I was reminded of Medea's anointing Jason* before the great episode of the iron bulls, but to-day William really could not be going up country to see a railroad for the first time. I knew this to be one of his great schemes, but he was not fitted to appear in public, or to front an observing world of strangers. As I appeared he essayed to rise, but Mrs Todd pushed him back into the chair.

'Set where you be till it dries on,' she insisted. 'Land sakes, you'd think he'd get over bein' a boy some time or 'nother, gettin' along in years as he is. An' you'd think he'd seen full enough o' fish, but once a year he has to break loose like this, an' travel off way up back o' the Bowden place—far out o' my beat, 't is—an' go a trout fishin'!'

Her tone of amused scorn was so full of challenge that William changed color even under the green streaks.

'I want some change,' he said, looking at me and not at her. ' 'T is the prettiest little shady brook you ever saw.'

'If he ever fetched home more 'n a couple o' minnies,* 't would seem worth while,' Mrs Todd concluded, putting a last dab of the mysterious compound so perilously near her brother's mouth that William flushed again and was silent.

A little later I witnessed his escape, when Mrs Todd had taken the foolish risk of going down cellar. There was a horse and wagon outside the garden fence, and presently we stood where we could see him driving up the hill with thoughtless speed. Mrs Todd said nothing, but watched him affectionately out of sight.

'It serves to keep the mosquitoes off,' she said, and a moment later it occurred to my slow mind that she spoke of the pennyroyal lotion. 'I don't know sometimes but William's kind of poetical,' she continued, in her gentlest voice. 'You'd think if anything could cure him of it, 't would be the fish business.'

It was only twenty minutes past six on a summer morning, but we both sat down to rest as if the activities of the day were over. Mrs Todd rocked gently for a time, and seemed to be lost, though not poorly, like Macbeth,* in her thoughts. At last she resumed relations with her actual surroundings. 'I shall now put my lobsters on. They'll make us a good supper,' she announced. 'Then I can let the fire out for all day; give it a holiday, same 's William. You can have a little one now, nice an' hot, if you ain't got all the breakfast you want. Yes, I'll put the lobsters on. William was very thoughtful to bring 'em over; William *is* thoughtful; if he only had a spark o' ambition, there be few could match him.'

This unusual concession was afforded a sympathetic listener from the depths of the kitchen closet. Mrs Todd was getting out her old iron lobster pot, and began to speak of prosaic affairs. I hoped that I should hear something more about her brother and their island life, and sat idly by the kitchen window looking at the morning glories that shaded it, believing that some flaw of wind might set Mrs Todd's mind on its former course. Then it occurred to me that she had spoken about our supper rather than our dinner, and I guessed that she might have some great scheme before her for the day.

When I had loitered for some time and there was no further word about William, and at last I was conscious of receiving no attention whatever, I went away. It was something of a disappointment to find that she put no hindrance in the way of my usual morning affairs, of going up to the empty little white schoolhouse on the hill where I did my task of writing. I had been almost sure of a holiday when I discovered that Mrs Todd was likely to take one herself; we had not been far afield to gather herbs and pleasures for many days now, but a little later she had silently vanished. I found my luncheon ready on the table in the little entry, wrapped in its shining old homespun napkin, and as if by way of special consolation, there was a stone bottle of Mrs Todd's best spruce beer, with a long piece of cod line wound round it by which it could be lowered for coolness into the deep schoolhouse well.

I walked away with a dull supply of writing-paper and these provisions, feeling like a reluctant child who hopes to be called back at every step. There was no relenting voice to be heard, and when I reached the schoolhouse, I found that I had left an open window and a swinging shutter the day before, and the sea wind that blew at evening had fluttered my poor sheaf of papers all about the room.

So the day did not begin very well, and I began to recognize that it was one of the days when nothing could be done without company. The truth was that my heart had gone trouting with William, but it would have been too selfish to say a word even to one's self about spoiling his day. If there is one way above another of getting so close to nature that one simply is a piece of nature, following a primeval instinct with perfect self-forgetfulness and forgetting everything except the dreamy consciousness of pleasant freedom, it is to take the course of a shady trout brook. The dark pools and the sunny shallows beckon one on; the wedge of sky between the trees on either bank, the speaking, companioning noise of the water, the amazing importance of what one is doing, and the constant sense of life and beauty make a strange transformation of the quick hours. I had a sudden memory of all this, and another, and another. I could not get myself free from 'fishing and wishing.'

At that moment I heard the unusual sound of wheels, and I looked past the high-growing thicket of wild-roses and straggling sumach to see the white nose and meagre shape of the Caplin horse; then I saw William sitting in the open wagon, with a small expectant smile upon his face.

'I've got two lines,' he said. 'I was quite a piece up the road. I thought perhaps 't was so you'd feel like going.'

There was enough excitement for most occasions in hearing William speak three sentences at once. Words seemed but vain to me at that bright moment. I stepped back from the schoolhouse window with a beating heart. The spruce-beer bottle was not yet in the well, and with that and my luncheon, and Pleasure at the helm, I went out into the happy world. The land breeze was blowing, and, as we turned away, I saw a flutter of white go past the window as I left the schoolhouse and my morning's work to their neglected fate.

II

ONE seldom gave way to a cruel impulse to look at an ancient seafaring William, but one felt as if he were a growing boy; I only hope that he felt much the same about me. He did not wear the fishing clothes that belonged to his sea-going life, but a strangely shaped old suit of tea-colored linen garments that might have been brought home years ago from Canton or Bombay. William had a peculiar way of giving silent assent when one spoke, but of answering your unspoken thoughts as if they reached him better than words. 'I find them very easy,' he said, frankly referring to the clothes. 'Father had them in his old sea-chest.'

The antique fashion, a quaint touch of foreign grace and even imagination about the cut were very pleasing; if ever Mr William Blackett had faintly resembled an old beau, it was upon that day. He now appeared to feel as if everything had been explained between us, as if everything were quite understood; and we drove for some distance without finding it necessary to speak again about anything. At last, when it must have been a little past nine o'clock, he stopped the horse beside a small farmhouse, and nodded when I asked if I should get down from the wagon. 'You can steer about northeast right across the pasture,' he said, looking from under the eaves of his hat with an expectant smile. 'I always leave the team here.'

I helped to unfasten the harness, and William led the horse away to the barn. It was a poor-looking little place, and a forlorn woman looked at us through the window before she appeared at the door. I told her that Mr Blackett and I came up from the Landing to go fishing. 'He keeps a-comin', don't he?' she answered, with a funny little laugh, to which I was at a loss to find answer. When he joined us, I could not see that he took notice of her presence in any way, except to take an armful of dried salt fish from a corded stack in the back of the wagon which had been carefully covered with a piece of old sail. We had left a wake of their pungent flavor behind us all the way. I wondered what was going to become of the rest of them and some fresh lobsters which were also disclosed to view, but he laid the present gift on the doorstep without a word, and a few minutes later, when I looked back as we crossed the pasture, the fish were being carried into the house.

I could n t see any signs of a trout brook until I came close upon it in the bushy pasture, and presently we struck into the low woods of straggling spruce and fir mixed into a tangle of swamp maples and alders which stretched away on either hand up and down stream. We found an open place in the pasture where some taller trees seemed to have been over-looked rather than spared. The sun was bright and hot by this time, and I sat down in the shade while William produced his lines and cut and trimmed us each a slender rod. I wondered where Mrs Todd was spending the morning, and if later she would think that pirates had landed and captured me from the schoolhouse.

III

THE brook was giving that live, persistent call to a listener that trout brooks always make; it ran with a free, swift current even here, where it crossed an apparently level piece of land. I saw two unpromising, quick barbel chase each other upstream from bank to bank as we solemnly arranged our hooks and sinkers. I felt that William's glances changed from anxiety to relief when he found that I was used to such gear;

perhaps he felt that we must stay together if I could not bait my own hook, but we parted happily, full of a pleasing sense of companionship.

William had pointed me up the brook, but I chose to go down, which was only fair because it was his day, though one likes as well to follow and see where a brook goes as to find one's way to the places it comes from, and its tiny springs and headwaters, and in this case trout were not to be considered. William's only real anxiety was lest I might suffer from mosquitoes. His own complexion was still strangely impaired by its defenses, but I kept forgetting it, and looking to see if we were treading fresh pennyroyal underfoot, so efficient was Mrs Todd's remedy. I was conscious, after we parted, and I turned to see if he were already fishing, and saw him wave his hand gallantly as he went away, that our friendship had made a great gain.

The moment that I began to fish the brook, I had a sense of its emptiness; when my bait first touched the water and went lightly down the quick stream, I knew that there was nothing to lie in wait for it. It is the same certainty that comes when one knocks at the door of an empty house, a lack of answering consciousness and of possible response; it is quite different if there is any life within. But it was a lovely brook, and I went a long way through woods and breezy open pastures, and found a forsaken house and overgrown farm, and laid up many pleasures for future joy and remembrance. At the end of the morning I came back to our meeting-place hungry and without any fish. William was already waiting, and we did not mention the matter of trout. We ate our luncheons with good appetites, and William brought our two stone bottles of spruce beer from the deep place in the brook where he had left them to cool. Then we sat awhile longer in peace and quietness on the green banks.

As for William, he looked more boyish than ever, and kept a more remote and juvenile sort of silence. Once I wondered how he had come to be so curiously wrinkled, forgetting, absent-mindedly, to recognize the effects of time. He did not expect any one else to keep up a vain show of conversation, and so I was silent as well as he. I glanced at him now and then, but I watched the leaves tossing against the sky and the red cattle moving in the pasture. 'I don't know's we need head for home. It's early yet,' he said at last, and I was as startled as if one of the gray firs had spoken.

'I guess I'll go up-along and ask after Thankful Hight's folks,' he continued. 'Mother'd like to get word;' and I nodded a pleased assent.

IV

WILLIAM led the way across the pasture, and I followed with a deep sense of pleased anticipation. I do not believe that my companion had expected me to make any objection, but I knew that he was gratified by the easy way that his plans for the day were being seconded. He gave a look at the sky to see if there were any portents, but the sky was frankly blue; even the doubtful morning haze had disappeared.

We went northward along a rough, clayey road, across a bare-looking, sunburnt country full of tiresome long slopes where the sun was hot and bright, and I could not help observing the forlorn look of the farms. There was a great deal of pasture, but it looked deserted, and I wondered afresh why the people did not raise more sheep when that seemed the only possible use to make of their land. I said so to Mr Blackett, who gave me a look of pleased surprise.

'That's what She always maintains,' he said eagerly. 'She's right about it, too; well, you'll see!' I was glad to find myself approved, but I had not the least idea whom he meant, and waited until he felt like speaking again.

A few minutes later we drove down a steep hill and entered a large tract of dark spruce woods. It was delightful to be sheltered from the afternoon sun, and when we had gone some distance in the shade, to my great pleasure William turned the horse's head toward some bars, which he let down, and I drove through into one of those narrow, still, sweet-scented by-ways which seem to be paths rather than roads. Often we had to put aside the heavy drooping branches which barred the way, and once, when a sharp twig struck William in

the face, he announced with such spirit that somebody ought
to go through there with an axe, that I felt unexpectedly
guilty. So far as I now remember, this was William's only
remark all the way through the woods to Thankful Hight's
folks, but from time to time he pointed or nodded at something which I might have missed: a sleepy little owl snuggled
into the bend of a branch, or a tall stalk of cardinal flowers
where the sunlight came down at the edge of a small, bright
piece of marsh. Many times, being used to the company of
Mrs Todd and other friends who were in the habit of talking,
I came near making an idle remark to William, but I was for
the most part happily preserved; to be with him only for a
short time was to live on a different level, where thoughts
served best because they were thoughts in common; the pri-
mary effect upon our minds of the simple things and beauties
that we saw. Once when I caught sight of a lovely gay pigeon-
woodpecker eyeing us curiously from a dead branch, and
instinctively turned toward William, he gave an indulgent,
comprehending nod which silenced me all the rest of the way.
The wood-road was not a place for common noisy conversation; one would interrupt the birds and all the still little beasts
that belonged there. But it was mortifying to find how strong
the habit of idle speech may become in one's self. One need
not always be saying something in this noisy world. I grew
conscious of the difference between William's usual fashion
of life and mine; for him there were long days of silence in a
sea-going boat, and I could believe that he and his mother
usually spoke very little because they so perfectly understood
each other. There was something peculiarly unresponding
about their quiet island in the sea, solidly fixed into the still
foundations of the world, against whose rocky shores the sea
beats and calls and is unanswered.

We were quite half an hour going through the woods; the
horse's feet made no sound on the brown, soft track under
the dark evergreens. I thought that we should come out at last
into more pastures, but there was no half-wooded strip of land
at the end; the high woods grew squarely against an old stone
wall and a sunshiny open field, and we came out suddenly
into broad daylight that startled us and even startled the

horse, who might have been napping as he walked, like an old soldier. The field sloped up to a low unpainted house that faced the east. Behind it were long, frost-whitened ledges that made the hill, with strips of green turf and bushes between. It was the wildest, most Titanic sort of pasture country up there; there was a sort of daring in putting a frail wooden house before it, though it might have the homely field and honest woods to front against. You thought of the elements and even of possible volcanoes as you looked up the stony heights. Suddenly I saw that a region of what I had thought gray stones was slowly moving, as if the sun was making my eyesight unsteady.

'There's the sheep!' exclaimed William, pointing eagerly. 'You see the sheep?' and sure enough, it was a great company of woolly backs, which seemed to have taken a mysterious protective resemblance to the ledges themselves. I could discover but little chance for pasturage on that high sunburnt ridge, but the sheep were moving steadily in a satisfied way as they fed along the slopes and hollows.

'I never have seen half so many sheep as these, all summer long!' I cried with admiration.

'There ain't so many,' answered William soberly. 'It's a great sight. They do so well because they're shepherded, but you can't beat sense into some folks.'

'You mean that somebody stays and watches them?' I asked.

'She observed years ago in her readin' that they don't turn out their flocks without protection anywhere but in the State o' Maine,' returned William. 'First thing that put it into her mind was a little old book mother's got; she read it one time when she come out to the Island. They call it the "Shepherd o' Salisbury Plain."* 'T wasn't the purpose o' the book to most, but when she read it, "There, Mis' Blackett!" she said, "that's where we've all lacked sense; our Bibles ought to have taught us that what sheep need is a shepherd."* You see most folks about here gave up sheep-raisin' years ago 'count o' the dogs. So she gave up school-teachin' and went out to tend her flock, and has shepherded ever since, an' done well.'

For William, this approached an oration. He spoke with

enthusiasm, and I shared the triumph of the moment. 'There she is now!' he exclaimed, in a different tone, as the tall figure of a woman came following the flock and stood still on the ridge, looking toward us as if her eyes had been quick to see a strange object in the familiar emptiness of the field. William stood up in the wagon, and I thought he was going to call or wave his hand to her, but he sat down again more clumsily than if the wagon had made the familiar motion of a boat, and we drove on toward the house.

It was a most solitary place to live,—a place where one might think that a life could hide itself. The thick woods were between the farm and the main road, and as one looked up and down the country, there was no other house in sight.

'Potatoes look well,' announced William. 'The old folks used to say that there wa'n't no better land outdoors than the Hight field.'

I found myself possessed of a surprising interest in the shepherdess, who stood far away in the hill pasture with her great flock, like a figure of Millet's,* high against the sky.

V

EVERYTHING about the old farmhouse was clean and orderly, as if the green dooryard were not only swept, but dusted. I saw a flock of turkeys stepping off carefully at a distance, but there was not the usual untidy flock of hens about the place to make everything look in disarray. William helped me out of the wagon as carefully as if I had been his mother, and nodded toward the open door with a reassuring look at me; but I waited until he had tied the horse and could lead the way, himself. He took off his hat just as we were going in, and stopped for a moment to smooth his thin gray hair with his hand, by which I saw that we had an affair of some ceremony. We entered an old-fashioned country kitchen, the floor scrubbed into unevenness, and the doors well polished by the touch of hands. In a large chair facing the window there sat a masterful-looking old woman with the features of a warlike

Roman emperor, emphasized by a bonnet-like black cap with a band of green ribbon. Her sceptre was a palm-leaf fan.

William crossed the room toward her, and bent his head close to her ear.

'Feelin' pretty well to-day, Mis' Hight?' he asked, with all the voice his narrow chest could muster.

'No, I ain't, William. Here I have to set,' she answered coldly, but she gave an inquiring glance over his shoulder at me.

'This is the young lady who is stopping with Almiry this summer,' he explained, and I approached as if to give the countersign. She offered her left hand with considerable dignity, but her expression never seemed to change for the better. A moment later she said that she was pleased to meet me, and I felt as if the worst were over. William must have felt some apprehension, while I was only ignorant, as we had come across the field. Our hostess was more than disapproving, she was forbidding; but I was not long in suspecting that she felt the natural resentment of a strong energy that has been defeated by illness and made the spoil of captivity.

'Mother well as usual since you was up last year?' and William replied by a series of cheerful nods. The mention of dear Mrs Blackett was a help to any conversation.

'Been fishin', ashore,' he explained, in a somewhat conciliatory voice. 'Thought you'd like a few for winter,' which explained at once the generous freight we had brought in the back of the wagon. I could see that the offering was no surprise, and that Mrs Hight was interested.

'Well, I expect they're good as the last,' she said, but did not even approach a smile. She kept a straight, discerning eye upon me.

'Give the lady a cheer,' she admonished William, who hastened to place close by her side one of the straight-backed chairs that stood against the kitchen wall. Then he lingered for a moment like a timid boy. I could see that he wore a look of resolve, but he did not ask the permission for which he evidently waited.

'You can go search for Esther,' she said, at the end of a long pause that became anxious for both her guests. 'Esther'd like to see her;' and William in his pale nankeens disappeared with one light step and was off.

VI

'DON'T speak too loud, it jars a person's head,' directed Mrs Hight plainly. 'Clear an' distinct is what reaches me best. Any news to the Landin'?'

I was happily furnished with the particulars of a sudden death, and an engagement of marriage between a Caplin, a seafaring widower home from his voyage, and one of the younger Harrises; and now Mrs Hight really smiled and settled herself in her chair. We exhausted one subject completely before we turned to the other. One of the returning turkeys took an unwarrantable liberty, and, mounting the doorstep, came in and walked about the kitchen without being observed by its strict owner; and the tin dipper slipped off its nail behind us and made an astonishing noise, and jar enough to reach Mrs Hight's inner ear and make her turn her head to look at it; but we talked straight on. We came at last to understand each other upon such terms of friendship that she unbent her majestic port and complained to me as any poor old woman might of the hardships of her illness. She had already fixed various dates upon the sad certainty of the year when she had the shock, which had left her perfectly helpless except for a clumsy left hand which fanned and gestured, and settled and resettled the folds of her dress, but could do no comfortable time-shortening work.

'Yes 'm, you can feel sure I use it what I can,' she said severely. ' 'T was a long spell before I could let Esther go forth in the mornin' till she'd got me up an' dressed me, but now she leaves things ready overnight and I get 'em as I want 'em with my light pair o' tongs, and I feel very able about helpin' myself to what I once did. Then when Esther returns, all she has to do is to push me out here into the kitchen. Some parts o' the year Esther stays out all night, them moonlight nights when the dogs are apt to be after the sheep, but she don't use herself as hard as she once had to. She's well able to hire somebody, Esther is, but there, you can't find no hired man that wants to git up before five o'clock nowadays; 't ain't as 't was in my time. They're liable to fall asleep, too, and them moonlight nights she's so anxious she can't sleep, and out she goes. There's a kind of a fold, she calls it, up there in a sheltered spot, and she sleeps up in a little shed she's got,— built it herself for lambin' time and when the poor foolish creatur's gets hurt or anything. I've never seen it, but she says it's in a lovely spot and always pleasant in any weather. You see off, other side of the ridge, to the south'ard, where there's houses. I used to think some time I'd get up to see it again, and all them spots she lives in, but I sha'n't now. I'm beginnin' to go back; an' 't ain't surprisin'. I've kind of got used to disappointments,' and the poor soul drew a deep sigh.

VII

IT was long before we noticed the lapse of time; I not only told every circumstance known to me of recent events among the households of Mrs Todd's neighborhood at the shore, but Mrs Hight became more and more communicative on her part, and went carefully into the genealogical descent and personal experience of many acquaintances, until between us we had pretty nearly circumnavigated the globe and reached Dunnet Landing from an opposite direction to that in which we had started. It was long before my own interest began to flag; there was a flavor of the best sort in her definite and descriptive fashion of speech. It may be only a fancy of my own that in the sound and value of many words, with their lengthened vowels and doubled cadences, there is some faint survival on the Maine coast of the sound of English speech of Chaucer's time.*

At last Mrs Thankful Hight gave a suspicious look through the window.

'Where do you suppose they be?' she asked me. 'Esther must ha' been off to the far edge o' everything. I doubt

William ain't been able to find her; can't he hear their bells? His hearin' all right?'

William had heard some herons that morning which were beyond the reach of my own ears, and almost beyond eyesight in the upper skies, and I told her so. I was luckily preserved by some unconscious instinct from saying that we had seen the shepherdess so near as we crossed the field. Unless she had fled faster than Atalanta,* William must have been but a few minutes in reaching her immediate neighborhood. I now discovered with a quick leap of amusement and delight in my heart that I had fallen upon a serious chapter of romance. The old woman looked suspiciously at me, and I made a dash to cover with a new piece of information; but she listened with lofty indifference, and soon interrupted my eager statements.

'Ain't William been gone some considerable time?' she demanded, and then in a milder tone: 'The time has re'lly flown; I do enjoy havin' company. I set here alone a sight o' long days. Sheep is dreadful fools; I expect they heard a strange step, and set right off through bush an' brier, spite of all she could do. But William might have the sense to return, 'stead o' searchin' about. I want to inquire of him about his mother. What was you goin' to say? I guess you'll have time to relate it.'

My powers of entertainment were on the ebb, but I doubled my diligence and we went on for another half-hour at least with banners flying, but still William did not reappear. Mrs Hight frankly began to show fatigue.

'Somethin' 's happened, an' he's stopped to help her,' groaned the old lady, in the middle of what I had found to tell her about a rumor of disaffection with the minister of a town I merely knew by name in the weekly newspaper to which Mrs Todd subscribed. 'You step to the door, dear, an' look if you can't see 'em.' I promptly stepped, and once outside the house I looked anxiously in the direction which William had taken.

To my astonishment I saw all the sheep so near that I wonder we had not been aware in the house of every bleat and tinkle. And there, within a stone's-throw, on the first long gray ledge that showed above the juniper, were William and the shepherdess engaged in pleasant conversation. At first I was provoked and then amused, and a thrill of sympathy warmed my whole heart. They had seen me and risen as if by magic; I had a sense of being the messenger of Fate. One could almost hear their sighs of regret as I appeared; they must have passed a lovely afternoon. I hurried into the house with the reassuring news that they were not only in sight but perfectly safe, with all the sheep.

VIII

MRS HIGHT, like myself, was spent with conversation, and had ceased even the one activity of fanning herself. I brought a desired drink of water, and happily remembered some fruit that was left from my luncheon. She revived with splendid vigor, and told me the simple history of her later years since she had been smitten in the prime of her life by the stroke of paralysis, and her husband had died and left her alone with Esther and a mortgage on their farm. There was only one field of good land, but they owned a great region of sheep pasture and a little woodland. Esther had always been laughed at for her belief in sheep-raising when one by one their neighbors were giving up their flocks, and when everything had come to the point of despair she had raised all the money and bought all the sheep she could, insisting that Maine lambs were as good as any, and that there was a straight path by sea to Boston market. And by tending her flock herself she had managed to succeed; she had made money enough to pay off the mortgage five years ago, and now what they did not spend was safe in the bank. 'It has been stubborn work, day and night, summer and winter, an' now she's beginnin' to get along in years,' said the old mother sadly. 'She's tended me 'long o' the sheep, an' she's been a good girl right along, but she ought to have been a teacher;' and Mrs Hight sighed heavily and plied the fan again.

We heard voices, and William and Esther entered; they did not know that it was so late in the afternoon. William looked almost bold, and oddly like a happy young man rather than an ancient boy. As for Esther, she might have been Jeanne d'Arc* returned to her sheep, touched with age and gray with the ashes of a great remembrance. She wore the simple look of sainthood and unfeigned devotion. My heart was moved by the sight of her plain sweet face, weatherworn and gentle in its looks, her thin figure in its close dress, and the strong hand that clasped a shepherd's staff, and I could only hold William in new reverence; this silent farmer-fisherman who knew, and he alone, the noble and patient heart that beat within her breast. I am not sure that they acknowledged even to themselves that they had always been lovers; they could not consent to anything so definite or pronounced; but they were happy in being together in the world. Esther was untouched by the fret and fury of life; she had lived in sunshine and rain among her silly sheep, and been refined instead of coarsened, while her touching patience with a ramping old mother, stung by the sense of defeat and mourning her lost activities, had given back a lovely self-possession, and habit of sweet temper. I had seen enough of old Mrs Hight to know that nothing a sheep might do could vex a person who was used to the uncertainties and severities of her companionship.

IX

MRS HIGHT told her daughter at once that she had enjoyed a beautiful call, and got a great many new things to think of. This was said so frankly in my hearing that it gave a consciousness of high reward, and I was indeed recompensed by the grateful look in Esther's eyes. We did not speak much together, but we understood each other. For the poor old woman did not read, and could not sew or knit with her helpless hand, and they were far from any neighbors, while her spirit was as eager in age as in youth, and expected even more from a disappointing world. She had lived to see the mortgage paid and money in the bank, and Esther's success acknowledged on every hand, and there were still a few pleasures left in life. William had his mother, and Esther had hers, and they had not seen each other for a year, though Mrs Hight had spoken of a year's making no change in William even at his age. She must have been in the far eighties herself, but of a noble courage and persistence in the world she ruled from her stiff-backed rocking-chair.

William unloaded his gift of dried fish, each one chosen with perfect care, and Esther stood by, watching him, and then she walked across the field with us beside the wagon. I believed that I was the only one who knew their happy secret, and she blushed a little as we said good-by.

'I hope you ain't goin' to feel too tired, mother's so deaf; no, I hope you won't be tired,' she said kindly, speaking as if she well knew what tiredness was. We could hear the neglected sheep bleating on the hill in the next moment's silence. Then she smiled at me, a smile of noble patience, of uncomprehended sacrifice, which I can never forget. There was all the remembrance of disappointed hopes, the hardships of winter, the loneliness of single-handedness in her look, but I understood, and I love to remember her worn face and her young blue eyes.

'Good-by, William,' she said gently, and William said good-by, and gave her a quick glance, but he did not turn to look back, though I did, and waved my hand as she was putting up the bars behind us. Nor did he speak again until we had passed through the dark woods and were on our way homeward by the main road. The grave yearly visit had been changed from a hope into a happy memory.

'You can see the sea from the top of her pasture hill,' said William at last.

'Can you?' I asked, with surprise.

'Yes, it's very high land; the ledges up there show very plain in clear weather from the top of our island, and there's a high upstandin' tree that makes a landmark for the fishin' grounds.' And William gave a happy sigh.

When we had nearly reached the Landing, my companion looked over into the back of the wagon and saw that the piece of sailcloth was safe, with which he had covered the dried fish. 'I wish we had got some trout,' he said wistfully. 'They always appease Almiry, and make her feel't was worth while to go.'

I stole a glance at William Blackett. We had not seen a solitary mosquito, but there was a dark stripe across his mild face, which might have been an old scar won long ago in battle.

THE FOREIGNER

I

ONE evening, at the end of August, in Dunnet Landing, I heard Mrs Todd's firm footstep crossing the small front entry outside my door, and her conventional cough which served as a herald's trumpet, or a plain New England knock, in the harmony of our fellowship.

'Oh, please come in!' I cried, for it had been so still in the house that I supposed my friend and hostess had gone to see one of her neighbors. The first cold northeasterly storm of the season was blowing hard outside. Now and then there was a dash of great raindrops and a flick of wet lilac leaves against the window, but I could hear that the sea was already stirred to its dark depths, and the great rollers were coming in heavily against the shore. One might well believe that Summer was coming to a sad end that night, in the darkness and rain and sudden access of autumnal cold. It seemed as if there must be danger offshore among the outer islands.

'Oh, there!' exclaimed Mrs Todd, as she entered. 'I know nothing ain't ever happened out to Green Island since the world began, but I always do worry about mother in these great gales. You know those tidal waves occur sometimes down to the West Indies, and I get dwellin' on 'em so I can't set still in my chair, nor knit a common row to a stocking. William might get mooning, out in his small bo't, and not observe how the sea was making, an' meet with some accident. Yes, I thought I'd come in and set with you if you wa'n't busy. No, I never feel any concern about 'em in winter 'cause then they're prepared, and all ashore and everything snug. William ought to keep help, as I tell him; yes, he ought to keep help.'

I hastened to reassure my anxious guest by saying that Elijah Tilley had told me in the afternoon, when I came along the shore past the fish houses, that Johnny Bowden and the Captain were out at Green Island; he had seen them beating

up the bay, and thought they must have put into Burnt Island
cove, but one of the lobstermen brought word later that he
saw them hauling out at Green Island as he came by, and
Captain Bowden pointed ashore and shook his head to say
that he did not mean to try to get in. 'The old Miranda just
managed it, but she will have to stay at home a day or two and
put new patches in her sail,' I ended, not without pride in so
much circumstantial evidence.

Mrs Todd was alert in a moment. 'Then they'll all have a
very pleasant evening,' she assured me, apparently dismissing
all fears of tidal waves and other sea-going disasters. 'I
was urging Alick Bowden to go ashore some day and see
mother before cold weather. He's her own nephew; she sets a
great deal by him. And Johnny's a great chum o' William's;
don't you know the first day we had Johnny out 'long of us, he
took an' give William his money to keep for him that
he'd been a-savin', and William showed it to me an' was so
affected I thought he was goin' to shed tears? 'T was a dollar
an' eighty cents; yes, they'll have a beautiful evenin' all
together, and like 's not the sea 'll be flat as a doorstep come
morning.'

I had drawn a large wooden rocking-chair before the
fire, and Mrs Todd was sitting there jogging herself a little,
knitting fast, and wonderfully placid of countenance. There
came a fresh gust of wind and rain, and we could feel the
small wooden house rock and hear it creak as if it were a ship
at sea.

'Lord, hear the great breakers!' exclaimed Mrs Todd. 'How
they pound!—there, there! I always run of an idea that the
sea knows anger these nights and gets full o' fight. I can hear
the rote* o' them old black ledges way down the thorough-
fare. Calls up all those stormy verses in the Book o' Psalms;
David he knew how old sea-goin' folks have to quake at the
heart.'

I thought as I had never thought before of such anxieties.
The families of sailors and coastwise adventurers by sea must
always be worrying about somebody, this side of the world or
the other. There was hardly one of Mrs Todd's elder acquaintances, men or women, who had not at some time or other

made a sea voyage, and there was often no news until the
voyagers themselves came back to bring it.

'There's a roaring high overhead, and a roaring in the
deep sea,' said Mrs Todd solemnly, 'and they battle together
nights like this. No, I couldn't sleep; some women folks always
goes right to bed an' to sleep, so 's to forget, but 't ain't my
way. Well, it's a blessin' we don't all feel alike; there's hardly
any of our folks at sea to worry about, nowadays, but I can't
help my feelin's, an' I got thinking of mother all alone, if
William had happened to be out lobsterin' and couldn't
make the cove gettin' back.'

'They will have a pleasant evening,' I repeated. 'Captain
Bowden is the best of good company.'

'Mother'll make him some pancakes for his supper, like 's
not,' said Mrs Todd, clicking her knitting needles and giving
a pull at her yarn. Just then the old cat pushed open the
unlatched door and came straight toward her mistress's lap.
She was regarded severely as she stepped about and turned
on the broad expanse, and then made herself into a round
cushion of fur, but was not openly admonished. There was
another great blast of wind overhead, and a puff of smoke
came down the chimney.

'This makes me think o' the night Mis' Cap'n Tolland
died,' said Mrs Todd, half to herself. 'Folks used to say these
gales only blew when somebody's a-dyin', or the devil was a-
comin' for his own, but the worst man I ever knew died a real
pretty mornin' in June.'

'You have never told me any ghost stories,' said I; and such
was the gloomy weather and the influence of the night that I
was instantly filled with reluctance to have this suggestion
followed. I had not chosen the best of moments; just before I
spoke we had begun to feel as cheerful as possible. Mrs Todd
glanced doubtfully at the cat and then at me, with a strange
absent look, and I was really afraid that she was going to tell
me something that would haunt my thoughts on every dark
stormy night as long as I lived.

'Never mind now; tell me to-morrow by daylight, Mrs
Todd,' I hastened to say, but she still looked at me full of
doubt and deliberation.

'Ghost stories!' she answered. 'Yes, I don't know but I've heard a plenty of 'em first an' last. I was just sayin' to myself that this is like the night Mis' Cap'n Tolland died. 'T was the great line storm* in September all of thirty, or maybe forty, year ago. I ain't one that keeps much account o' time.'

'Tolland? That's a name I have never heard in Dunnet,' I said.

'Then you haven't looked well about the old part o' the buryin' ground, no'theast corner,' replied Mrs Todd. 'All their women folks lies there; the sea 's got most o' the men. They were a known family o' shipmasters in early times. Mother had a mate, Ellen Tolland, that she mourns to this day; died right in her bloom with quick consumption, but the rest o' that family was all boys but one, and older than she, an' they lived hard seafarin' lives an' all died hard. They were called very smart seamen. I've heard that when the youngest went into one o' the old shippin' houses in Boston, the head o' the firm called out to him: 'Did you say Tolland from Dunnet? That's recommendation enough for any vessel!' There was some o' them old shipmasters as tough as iron, an' they had the name o' usin' their crews very severe, but there wa'n't a man that wouldn't rather sign with 'em an' take his chances, than with the slack ones that didn't know how to meet accidents.'

II

THERE was so long a pause, and Mrs Todd still looked so absent-minded, that I was afraid she and the cat were growing drowsy together before the fire, and I should have no reminiscences at all. The wind struck the house again, so that we both started in our chairs and Mrs Todd gave a curious, startled look at me. The cat lifted her head and listened too, in the silence that followed, while after the wind sank we were more conscious than ever of the awful roar of the sea. The house jarred now and then, in a strange, disturbing way.

'Yes, they'll have a beautiful evening out to the island,' said

Mrs Todd again; but she did not say it gayly. I had not seen her before in her weaker moments.

'Who was Mrs Captain Tolland?' I asked eagerly, to change the current of our thoughts.

'I never knew her maiden name; if I ever heard it, I've gone an' forgot; 't would mean nothing to me,' answered Mrs Todd.

'She was a foreigner, an' he met with her out in the Island o' Jamaica. They said she'd been left a widow with property. Land knows what become of it; she was French born, an' her first husband was a Portugee, or somethin'.'

I kept silence now, a poor and insufficient question being worse than none.

'Cap'n John Tolland was the least smartest of any of 'em, but he was full smart enough, an' commanded a good brig at the time, in the sugar trade; he'd taken out a cargo o' pine lumber to the islands from somewheres up the river, an' had been loadin' for home in the port o' Kingston, an' had gone ashore that afternoon for his papers, an' remained afterwards 'long of three friends o' his, all shipmasters. They was havin' their suppers together in a tavern; 't was late in the evenin' an' they was more lively than usual, an' felt boyish; and over opposite was another house full o' company, real bright and pleasant lookin', with a lot o' lights, an' they heard somebody singin' very pretty to a guitar. They wa'n't in no go-to-meetin' condition, an' one of 'em, he slapped the table an' said, "Le''s go over an' hear that lady sing!" an' over they all went, good honest sailors, but three sheets in the wind, and stepped in as if they was invited, an' made their bows inside the door, an' asked if they could hear the music; they were all respectable well-dressed men. They saw the woman that had the guitar, an' there was a company a-listenin', regular highbinders* all of 'em; an' there was a long table all spread out with big candlesticks like little trees o' light, and a sight o' glass an' silver ware; an' part o' the men was young officers in uniform, an' the colored folks was steppin' round servin' 'em, an' they had the lady singin'. 'T was a wasteful scene, an' a loud talkin' company, an' though they was three sheets in the wind them-selves there wa'n't one o' them cap'ns but had sense to perceive it. The others had pushed back their chairs, an' their decanters an' glasses was standin' thick about, an' they was teasin' the one that was singin' as if they'd just got her in to amuse 'em. But they quieted down; one o' the young officers had beautiful manners, an' invited the four cap'ns to join 'em, very polite; 't was a kind of public house, and after they'd all heard another song, he come to consult with 'em whether they wouldn't git up and dance a hornpipe or somethin' to the lady's music.

'They was all elderly men an' shipmasters, and owned property; two of 'em was church members in good standin',' continued Mrs Todd loftily, 'an' they wouldn't lend theirselves to no such kick-shows* as that, an' spite o' bein' three sheets in the wind, as I have once observed; they waved aside the tumblers of wine the young officer was pourin' out for 'em so freehanded, and said they should rather be excused. An' when they all rose, still very dignified, as I've been well informed, and made their partin' bows and was goin' out, them young sports got round 'em an' tried to prevent 'em, and they had to push an' strive considerable, but out they come. There was this Cap'n Tolland and two Cap'n Bowdens, and the fourth was my own father.' (Mrs Todd spoke slowly, as if to impress the value of her authority.) 'Two of them was very religious, upright men, but they would have their night off sometimes, all o' them old-fashioned cap'ns, when they was free of business and ready to leave port.

'An' they went back to their tavern an' got their bills paid, an' set down kind o' mad with everybody by the front windows, mistrusting some o' their tavern charges, like 's not, by that time, an' when they got tempered down, they watched the house over across, where the party was.

'There was a kind of a grove o' trees between the house an' the road, an' they heard the guitar a-goin' an' a-stoppin' short by turns, and pretty soon somebody began to screech, an' they saw a white dress come runnin' out through the bushes, an' tumbled over each other in their haste to offer help; an' out she come, with the guitar, cryin' into the street, and they just walked off four square with her amongst 'em, down toward the wharves where they felt more to home. They couldn't make out at first what 't was she spoke,—Cap'n Lorenzo Bowden was well acquainted in Havre an' Bordeaux, an' spoke a poor quality o' French, an' she knew a little mite o' English, but not much; and they come somehow or other to discern that she was in real distress. Her husband and her children had died o' yellow fever; they'd all come up to Kingston from one o' the far Wind'ard Islands to get passage on a steamer to France, an' a negro had stole their money off her husband while he lay sick o' the fever, an' she had been befriended some, but the folks that knew about her had died too; it had been a dreadful run o' the fever that season, an' she fell at last to playin' an' singin' for hire, and for what money they'd throw to her round them harbor houses.

' 'T was a real hard case, an' when them cap'ns made out about it, there wa'n't one that meant to take leave without helpin' of her. They was pretty mellow, an' whatever they might lack o' prudence they more 'n made up with charity: they didn't want to see nobody abused, an' she was sort of a pretty woman, an' they stopped in the street then an' there an' drew lots who should take her aboard, bein' all bound home. An' the lot fell to Cap'n Jonathan Bowden who did act discouraged; his vessel had but small accommodations, though he could stow a big freight, an' she was a dreadful slow sailer through bein' square as a box, an' his first wife, that was livin' then, was a dreadful jealous woman. He threw himself right onto the mercy o' Cap'n Tolland.'

Mrs Todd indulged herself for a short time in a season of calm reflection.

'I always thought they'd have done better, and more reasonable, to give her some money to pay her passage home to France, or wherever she may have wanted to go,' she continued.

I nodded and looked for the rest of the story.

'Father told mother,' said Mrs Todd confidentially, 'that Cap'n Jonathan Bowden an' Cap'n John Tolland had both taken a little more than usual; I wouldn't have you think, either, that they both wasn't the best o' men, an' they was solemn as owls, and argued the matter between 'em, an' waved aside the other two when they tried to put their oars in.

An' spite o' Cap'n Tolland's bein' a settled old bachelor they fixed it that he was to take the prize on his brig; she was a fast sailer, and there was a good spare cabin or two where he'd sometimes carried passengers, but he'd filled 'em with bags o' sugar on his own account an' was loaded very heavy beside. He said he'd shift the sugar an' get along somehow, an' the last the other three cap'ns saw of the party was Cap'n John handing the lady into his bo't, guitar and all, an' off they all set tow'ds their ships with their men rowin' 'em in the bright moonlight down to Port Royal where the anchorage was, an' where they all lay, goin' out with the tide an' mornin' wind at break o' day. An' the others thought they heard music of the guitar, two o' the bo'ts kept well together, but it may have come from another source.'

'Well; and then?' I asked eagerly after a pause. Mrs Todd was almost laughing aloud over her knitting and nodding emphatically. We had forgotten all about the noise of the wind and sea.

'Lord bless you! he come sailing into Portland with his sugar, all in good time, an' they stepped right afore a justice o' the peace, and Cap'n John Tolland come paradin' home to Dunnet Landin' a married man. He owned one o' them thin, narrow-lookin' houses with one room each side o' the front door, and two slim black spruces spindlin' up against the front windows to make it gloomy inside. There was no horse nor cattle of course, though he owned pasture land, an' you could see rifts o' light right through the barn as you drove by. And there was a good excellent kitchen, but his sister reigned over that; she had a right to two rooms, and took the kitchen an' a bedroom that led out of it; an' bein' given no rights in the kitchen had angered the cap'n so they weren't on no kind o' speakin' terms. He preferred his old brig for comfort, but now and then, between voyages, he'd come home for a few days, just to show he was master over his part o' the house, and show Eliza she couldn't commit no trespass.

'They stayed a little while; 't was pretty spring weather, an' I used to see Cap'n John rollin' by with his arms full o' bundles from the store, lookin' as pleased and important as a boy; an' then they went right off to sea again, an' was gone a good many months. Next time he left her to live there alone, after they'd stopped at home together some weeks, an' they said she suffered from bein' at sea, but some said that the owners wouldn't have a woman aboard. 'T was before father was lost on that last voyage of his, an' he and mother went up once or twice to see them. Father said there wa'n't a mite o' harm in her, but somehow or other a sight o' prejudice arose; it may have been caused by the remarks of Eliza an' her feelin's tow'ds her brother. Even my mother had no regard for Eliza Tolland. But mother asked the cap'n's wife to come with her one evenin' to a social circle that was down to the meetin'-house vestry, so she'd get acquainted a little, an' she appeared very pretty until they started to have some singin' to the melodeon. Mari' Harris an' one o' the younger Caplin girls undertook to sing a duet, an' they sort o' flatted, an' she put her hands right up to her ears, and give a little squeal, an' went quick as could be an' give 'em the right notes, for she could read the music like plain print, an' made 'em try it over again. She was real willin' an' pleasant, but that didn't suit, an' she made faces when they got it wrong. An' then there fell a dead calm, an' we was all settin' round prim as dishes, an' my mother, that never expects ill feelin', asked her if she wouldn't sing somethin', an' up she got,—poor creatur', it all seems so different to me now,—an' sung a lovely little song standin' in the floor; it seemed to have something gay about it that kept a-repeatin', an' nobody could help keepin' time, an' all of a sudden she looked round at the tables and caught up a tin plate that somebody'd fetched a Washin'ton pie* in, an' she begun to drum on it with her fingers like one o' them tambourines, an' went right on singin' faster an' faster, and next minute she begun to dance a little pretty dance between the verses, just as light and pleasant as a child. You couldn't help seein' how pretty 't was; we all got to trottin' a foot, an' some o' the men clapped their hands quite loud, a-keepin' time, 't was so catchin', an' seemed so natural to her. There wa'n't one of 'em but enjoyed it; she just tried to do her part, an' some urged her on, till she stopped with a little twirl of her skirts an' went to her place again by mother. And I can see mother now, reachin' over an' smilin' an' pattin' her hand.

'But next day there was an awful scandal goin' in the parish, an' Mari' Harris reproached my mother to her face, an' I never wanted to see her since, but I've had to a good many times. I said Mis' Tolland didn't intend no impropriety,—I reminded her of David's dancin' before the Lord;* but she said such a man as David never would have thought o' dancin' right there in the Orthodox* vestry, and she felt I spoke with irreverence.

'And next Sunday Mis' Tolland come walkin' into our meeting, but I must say she acted like a cat in a strange garret, and went right out down the aisle with her head in air, from the pew Deacon Caplin had showed her into. 'T was just in the beginning of the long prayer. I wish she'd stayed through, whatever her reasons were. Whether she'd expected somethin' different, or misunderstood some o' the pastor's remarks, or what 't was, I don't really feel able to explain, but she kind o' declared war, at least folks thought so, an' war 't was from that time. I see she was cryin', or had been, as she passed by me; perhaps bein' in meetin' was what had power to make her feel homesick and strange.

'Cap'n John Tolland was away fittin' out; that next week he come home to see her and say farewell. He was lost with his ship in the Straits of Malacca, and she lived there alone in the old house a few months longer till she died. He left her well off; 't was said he hid his money about the house and she knew where 't was. Oh, I expect you've heard that story told over an' over twenty times, since you've been here at the Landin'?'

'Never one word,' I insisted.

'It was a good while ago,' explained Mrs Todd, with reassurance. 'Yes, it all happened a great while ago.'

III

At this moment, with a sudden flaw of the wind, some wet twigs outside blew against the window panes and made a noise like a distressed creature trying to get in. I started with sudden fear, and so did the cat, but Mrs Todd knitted away and did not even look over her shoulder.

'She was a good-looking woman; yes, I always thought Mis' Tolland was good-looking, though she had, as was reasonable, a sort of foreign cast, and she spoke very broken English, no better than a child. She was always at work about her house, or settin' at a front window with her sewing; she was a beautiful hand to embroider. Sometimes, summer evenings, when the windows was open, she'd set an' drum on her guitar, but I don't know as I ever heard her sing but once after the cap'n went away. She appeared very happy about havin' him, and took on dreadful at partin' when he was down here on the wharf, going back to Portland by boat to take ship for that last v'y'ge. He acted kind of ashamed, Cap'n John did; folks about here ain't so much accustomed to show their feelings. The whistle had blown an' they was waitin' for him to get aboard, an' he was put to it to know what to do and treated her very affectionate in spite of all impatience; but mother happened to be there and she went an' spoke, and I remember what a comfort she seemed to be. Mis' Tolland clung to her then, and she wouldn't give a glance after the boat when it had started, though the captain was very eager a-wavin' to her. She wanted mother to come home with her an' wouldn't let go her hand, and mother had just come in to stop all night with me an' had plenty o' time ashore, which didn't always happen, so they walked off together, an' 't was some considerable time before she got back.

' "I want you to neighbor with that poor lonesome creatur'," says mother to me, lookin' reproachful. "She's a stranger in a strange land,"* says mother. "I want you to make her have a sense that somebody feels kind to her."

' "Why, since that time she flaunted out o' meetin', folks have felt she liked other ways better 'n our'n," says I. I was provoked, because I'd had a nice supper ready, an' mother'd let it wait so long 't was spoiled. "I hope you'll like your supper!" I told her. I was dreadful ashamed afterward of speakin' so to mother.

' "What consequence is my supper?" says she to me; mother can be very stern,—"or your comfort or mine, beside letting a foreign person an' a stranger feel so desolate; she's done the best a woman could do in her lonesome place, and she asks nothing of anybody except a little common kindness. Think if 't was you in a foreign land!"

'And mother set down to drink her tea, an' I set down humbled enough over by the wall to wait till she finished. An' I did think it all over, an' next day I never said nothin', but I put on my bonnet, and went to see Mis' Cap'n Tolland, if 't was only for mother's sake. 'T was about three quarters of a mile up the road here, beyond the schoolhouse. I forgot to tell you that the cap'n had bought out his sister's right at three or four times what 't was worth, to save trouble, so they'd got clear o' her, an' I went round into the side yard sort o' friendly an' sociable, rather than stop an' deal with the knocker an' the front door. It looked so pleasant an' pretty I was glad I come; she had set a little table for supper, though 't was still early, with a white cloth on it, right out under an old apple tree close by the house. I noticed 't was same as with me at home, there was only one plate. She was just coming out with a dish; you couldn't see the door nor the table from the road.

'In the few weeks she'd been there she'd got some bloomin' pinks an' other flowers next the doorstep. Somehow it looked as if she'd known how to make it homelike for the cap'n. She asked me to set down; she was very polite, but she looked very mournful, and I spoke of mother, an' she put down her dish and caught holt o' me with both hands an' said my mother was an angel. When I see the tears in her eyes 't was all right between us, and we were always friendly after that, and mother had us come out and make a little visit that summer; but she come a foreigner and she went a foreigner, and never was anything but a stranger among our folks. She taught me a sight o' things about herbs I never knew before nor since; she was well acquainted with the virtues o' plants. She'd act awful secret about some things too, an' used to work charms for herself sometimes, an' some o' the neighbors told to an' fro after she died that they knew enough not to provoke her, but 't was all nonsense; 't is the believin' in such things that causes 'em to be any harm, an' so I told 'em,' confided Mrs Todd contemptuously. 'That first night I stopped to tea with her she'd cooked some eggs with some herb or other sprinkled all through, and 't was she that first led me to discern mushrooms; an' she went right down on her knees in my garden here when she saw I had my different officious* herbs. Yes, 't was she that learned me the proper use o' parsley too; she was a beautiful cook.'

Mrs Todd stopped talking, and rose, putting the cat gently in the chair, while she went away to get another stick of apple-tree wood. It was not an evening when one wished to let the fire go down, and we had a splendid bank of bright coals. I had always wondered where Mrs Todd had got such an un-usual knowledge of cookery, of the varieties of mushrooms, and the use of sorrel as a vegetable, and other blessings of that sort. I had long ago learned that she could vary her omelettes like a child of France, which was indeed a surprise in Dunnet Landing.

IV

ALL these revelations were of the deepest interest, and I was ready with a question as soon as Mrs Todd came in and had well settled the fire and herself and the cat again.

'I wonder why she never went back to France, after she was left alone?'

'She come here from the French islands,' explained Mrs Todd. 'I asked her once about her folks, an' she said they were all dead; 't was the fever took 'em. She made this her home, lonesome as 't was; she told me she hadn't been in France since she was "so small," and measured me off a child o' six. She'd lived right out in the country before, so that part wa'n't unusual to her. Oh yes, there was something very strange about her, and she hadn't been brought up in high circles nor nothing o' that kind. I think she'd been really pleased to have the cap'n marry her an' give her a good home, after all she'd passed through, and leave her free with his money an' all that. An' she got over bein' so strange-looking to me after a while, but 't was a very singular ex-pression: she wore a fixed smile that wa'n't a smile; there wa'n't no light behind it, same 's a lamp can't shine if it ain't lit. I don't know just how to express it, 't was a sort of made countenance.'

One could not help thinking of Sir Philip Sidney's phrase,* 'A made countenance, between simpering and smiling.'

'She took it hard, havin' the captain go off on that last voyage,' Mrs Todd went on. 'She said somethin' told her when they was partin' that he would never come back. He was lucky to speak a home-bound ship this side o' the Cape o' Good Hope, an' got a chance to send her a letter, an' that cheered her up. You often felt as if you was dealin' with a child's mind, for all she had so much information that other folks hadn't. I was a sight younger than I be now, and she made me imagine new things, and I got interested watchin' her an' findin' out what she had to say, but you couldn't get to no affectionateness with her. I used to blame me some- times; we used to be real good comrades goin' off for an afternoon, but I never give her a kiss till the day she laid in her coffin and it come to my heart there wa'n't no one else to do it.'

'And Captain Tolland died,' I suggested after a while.

'Yes, the cap'n was lost,' said Mrs Todd, 'and of course word didn't come for a good while after it happened. The letter come from the owners to my uncle, Cap'n Lorenzo Bowden, who was in charge of Cap'n Tolland's affairs at home, and he come right up for me an' said I must go with him to the house. I had known what it was to be a widow, myself, for near a year, an' there was plenty o' widow women along this coast that the sea had made desolate, but I never saw a heart break as I did then.

' 'T was this way: we walked together along the road, me an' uncle Lorenzo. You know how it leads straight from just above the schoolhouse to the brook bridge, and their house was just this side o' the brook bridge on the left hand; the cellar's there now, and a couple or three good-sized gray birches growin' in it. And when we come near enough I saw that the best room, this way, where she most never set, was all lighted up, and the curtains up so that the light shone bright down the road, and as we walked, those lights would dazzle and dazzle in my eyes, and I could hear the guitar a-goin', an' she was singin'. She heard our steps with her quick ears and come running to the door with her eyes a-shinin', an' all that set look gone out of her face, an' begun to talk French, gay as a bird, an' shook hands and behaved very pretty an' girlish, sayin' 't was her fête day.* I didn't know what she meant then. And she had gone an' put a wreath o' flowers on her hair an' wore a handsome gold chain that the cap'n had given her; an' there she was, poor creatur', makin' believe have a party all alone in her best room; 't was prim enough to discourage a person, with too many chairs set close to the walls, just as the cap'n's mother had left it, but she had put sort o' long garlands on the walls, droopin' very graceful, and a sight of green boughs in the corners, till it looked lovely, and all lit up with a lot o' candles.'

'Oh dear!' I sighed. 'Oh, Mrs Todd, what did you do?'

'She beheld our countenances,' answered Mrs Todd solemnly. 'I expect they was telling everything plain enough, but Cap'n Lorenzo spoke the sad words to her as if he had been her father; and she wavered a minute and then over she went on the floor before we could catch hold of her, and then we tried to bring her to herself and failed, and at last we carried her upstairs, an' I told uncle to run down and put out the lights, and then go fast as he could for Mrs Begg, being very experienced in sickness, an' he so did. I got off her clothes and her poor wreath, and I cried as I done it. We both stayed there that night, and the doctor said 't was a shock when he come in the morning; he'd been over to Black Island an' had to stay all night with a very sick child.'

'You said that she lived alone some time after the news came,' I reminded Mrs Todd then.

'Oh yes, dear,' answered my friend sadly, 'but it wa'n't what you'd call livin'; no, it was only dyin', though at a snail's pace. She never went out again those few months, but for a while she could manage to get about the house a little, and do what was needed, an' I never let two days go by without seein' her or hearin' from her. She never took much notice as I came an' went except to answer if I asked her anything. Mother was the one who gave her the only comfort.'

'What was that?' I asked softly.

'She said that anybody in such trouble ought to see their minister, mother did, and one day she spoke to Mis' Tolland, and found that the poor soul had been believin' all the time that there weren't any priests here. We'd come to know she was a Catholic by her beads and all, and that had set some narrow minds against her. And mother explained it just as she would to a child; and uncle Lorenzo sent word right off somewheres up river by a packet that was bound up the bay, and the first o' the week a priest come by the boat, an' uncle Lorenzo was on the wharf 'tendin' to some business; so they just come up for me, and I walked with him to show him the house. He was a kind-hearted old man; he looked so benevolent an' fatherly I could ha' stopped an' told him my own troubles; yes, I was satisfied when I first saw his face, an' when poor Mis' Tolland beheld him enter the room, she went right down on her knees and clasped her hands together to him as if he'd come to save her life, and he lifted her up and blessed her, an' I left 'em together, and slipped out into the open field and walked there in sight so if they needed to call me, and I had my own thoughts. At last I saw him at the door; he had to catch the return boat. I meant to walk back with him and offer him some supper, but he said no, and said he was comin' again if needed, and signed me to go into the house to her, and shook his head in a way that meant he understood everything. I can see him now; he walked with a cane, rather tired and feeble; I wished somebody would come along, so 's to carry him down to the shore.

'Mis' Tolland looked up at me with a new look when I went in, an' she even took hold o' my hand and kept it. He had put some oil on her forehead, but nothing anybody could do would keep her alive very long; 't was his medicine for the soul rather 'n the body. I helped her to bed, and next morning she couldn't get up to dress her, and that was Monday, and she began to fail, and 't was Friday night she died.' (Mrs Todd spoke with unusual haste and lack of detail.) 'Mrs Begg and I watched with her, and made everything nice and proper, and after all the ill will there was a good number gathered to the funeral. 'T was in Reverend Mr Bascom's day, and he done very well in his prayer, considering he couldn't fill in with

mentioning all the near connections by name as was his habit.
He spoke very feeling about her being a stranger and twice
widowed, and all he said about her being reared among the
heathen was to observe that there might be roads leadin' up
to the New Jerusalem* from various points. I says to myself
that I guessed quite a number must ha' reached there that
wa'n't able to set out from Dunnet Landin'!'

Mrs Todd gave an odd little laugh as she bent toward the
firelight to pick up a dropped stitch in her knitting, and then
I heard a heartfelt sigh.

' 'T was most forty years ago,' she said; 'Most everybody's
gone a'ready that was there that day.'

V

SUDDENLY Mrs Todd gave an energetic shrug of her
shoulders, and a quick look at me, and I saw that the sails of
her narrative were filled with a fresh breeze.

'Uncle Lorenzo, Cap'n Bowden that I have referred to'—
'Certainly!' I agreed with eager expectation.

'He was the one that had been left in charge of Cap'n John
Tolland's affairs, and had now come to be of unforeseen
importance.

'Mrs Begg an' I had stayed in the house both before an'
after Mis' Tolland's decease, and she was now in haste to be
gone, having affairs to call her home; but uncle come to me
as the exercises was beginning, and said he thought I'd better
remain at the house while they went to the buryin' ground. I
couldn't understand his reasons, an' I felt disappointed, bein'
as near to her as most anybody; 't was rough weather, so
mother couldn't get in, and didn't even hear Mis' Tolland
was gone till next day. I just nodded to satisfy him, 't wa'n't no
time to discuss anything. Uncle seemed flustered; he'd gone
out deep-sea fishin' the day she died, and the storm I told you
of rose very sudden, so they got blown off way down the coast
beyond Monhegan, and he'd just got back in time to dress
himself and come.

'I set there in the house after I'd watched her away down

the straight road far 's I could see from the door; 't was a little short walkin' funeral an' a cloudy sky, so everything looked dull an' gray, an' it crawled along all in one piece, same 's walking funerals do, an' I wondered how it ever come to the Lord's mind to let her begin down among them gay islands all heat and sun, and end up here among the rocks with a north wind blowin'. 'T was a gale that begun the afternoon before she died, and had kept blowin' off an' on ever since. I'd thought more than once how glad I should be to get home an' out o' sound o' them black spruces a-beatin' an' scratchin' at the front windows.

'I set to work pretty soon to put the chairs back, an' set outdoors some that was borrowed, an' I went out in the kitchen, an' I made up a good fire in case somebody come an' wanted a cup o' tea; but I didn't expect any one to travel way back to the house unless 't was uncle Lorenzo. 'T was growin' so chilly that I fetched some kindlin' wood and made fires in both the fore rooms. Then I set down an' begun to feel as usual, and I got my knittin' out of a drawer. You can't be sorry for a poor creatur' that's come to the end o' all her troubles; my only discomfort was I thought I'd ought to feel worse at losin' her than I did; I was younger then than I be now. And as I set there, I begun to hear some long notes o' dronin' music from upstairs that chilled me to the bone.'

Mrs Todd gave a hasty glance at me.

'Quick 's I could gather me, I went right upstairs to see what 't was,' she added eagerly, 'an' 't was just what I might ha' known. She'd always kept her guitar hangin' right against the wall in her room; 't was tied by a blue ribbon, and there was a window left wide open; the wind was veerin' a good deal, an' it slanted in and searched the room. The strings was jarrin' yet.

' 'T was growin' pretty late in the afternoon, an' I begun to feel lonesome as I shouldn't now, and I was disappointed at having to stay there, the more I thought it over, but after a while I saw Cap'n Lorenzo polin' back up the road all alone, and when he come nearer I could see he had a bundle under his arm and had shifted his best black clothes for his every-day ones. I run out and put some tea into the teapot and set it back on the stove to draw, an' when he come in I reached down a little jug o' spirits,—Cap'n Tolland had left his house well provisioned as if his wife was goin' to put to sea same 's himself, an' there she'd gone an' left it. There was some cake that Mis' Begg an' I had made the day before. I thought that uncle an' me had a good right to the funeral supper, even if there wa'n't any one to join us. I was lookin' forward to my cup o' tea; 't was beautiful tea out of a green lacquered chest that I've got now.'

'You must have felt very tired,' said I, eagerly listening.

'I was 'most beat out, with watchin' an' tendin' and all,' answered Mrs Todd, with as much sympathy in her voice as if she were speaking of another person. 'But I called out to uncle as he came in, "Well, I expect it's all over now, an' we've all done what we could. I thought we'd better have some tea or somethin' before we go home. Come right out in the kitchen, sir," says I, never thinking but we only had to let the fires out and lock up everything safe an' eat our refreshment, an' go home.

' "I want both of us to stop here to-night," says uncle, looking at me very important.

' "Oh, what for?" says I, kind o' fretful.

' "I've got my proper reasons," says uncle. "I'll see you well satisfied, Almira. Your tongue ain't so easy-goin' as some o' the women folks, an' there's property here to take charge of that you don't know nothin' at all about."

' "What do you mean?" says I.

' "Cap'n Tolland acquainted me with his affairs; he hadn't no sort o' confidence in nobody but me an' his wife, after he was tricked into signin' that Portland note, an' lost money. An' she didn't know nothin' about business; but what he didn't take to sea to be sunk with him he's hid somewhere in this house. I expect Mis' Tolland may have told you where she kept things?" said uncle.

'I see he was dependin' a good deal on my answer,' said Mrs Todd, 'but I had to disappoint him; no, she had never said nothin' to me.

' "Well, then, we've got to make a search," says he, with considerable relish; but he was all tired and worked up, and we set down to the table, an' he had somethin', an' I took my desired cup o' tea, and then I begun to feel more interested.

' "Where you goin' to look first?" says I, but he give me a short look an' made no answer, and begun to mix me a very small portion out of the jug, in another glass. I took it to please him; he said I looked tired, speakin' real fatherly, and I did feel better for it, and we set talkin' a few minutes, an' then he started for the cellar, carrying an old ship's lantern he fetched out o' the stairway an' lit.

' "What are you lookin' for, some kind of a chist?" I inquired, and he said yes. All of a sudden it come to me to ask who was the heirs; Eliza Tolland, Cap'n John's own sister, had never demeaned herself to come near the funeral, and uncle Lorenzo faced right about and begun to laugh, sort o' pleased. I thought queer of it; 't wa'n't what he'd taken, which would be nothin' to an old weathered sailor like him.

' "Who's the heir?" says I the second time.

' "Why, it's *you*, Almiry,' says he; and I was so took aback I set right down on the turn o' the cellar stairs.

' "Yes 't is," said uncle Lorenzo. "I'm glad of it too. Some thought she didn't have no sense but foreign sense, an' a poor stock o' that, but she said you was friendly to her, an' one day after she got news of Tolland's death, an' I had fetched up his will that left everything to her, she said she was goin' to make a writin', so 's you could have things after she was gone, an' she give five hundred to me for bein' executor. Square Pease fixed up the paper, an' she signed it; it's all accordin' to law." There, I begun to cry,' said Mrs Todd; 'I couldn't help it. I wished I had her back again to do somethin' for, an' to make her know I felt sisterly to her more 'n I'd ever showed, an' it come over me 't was all too late, an' I cried the more, till uncle showed impatience, an' I got up an' stumbled along down cellar with my apern to my eyes the greater part of the time.

' "I'm goin' to have a clean search," says he; "you hold the light." An' I held it, and he rummaged in the arches an' under the stairs, an' over in some old closet where he reached out bottles an' stone jugs an' canted some kags an' one or two casks, an' chuckled well when he heard there was somethin' inside,—but there wa'n't nothin' to find but things usual in a cellar, an' then the old lantern was givin' out an' we come away.

' "He spoke to me of a chist, Cap'n Tolland did," says uncle in a whisper. "He said a good sound chist was as safe a bank as there was, an' I beat him out of such nonsense, 'count o' fire an' other risks." "There's no chist in the rooms above," says I; "no, uncle, there ain't no sea-chist, for I've been here long enough to see what there was to be seen." Yet he wouldn't feel contented till he'd mounted up into the toploft; 't was one o' them single, hip-roofed houses that don't give proper accommodation for a real garret, like Cap'n Littlepage's down here at the Landin'. There was broken furniture and rubbish, an' he let down a terrible sight o' dust into the front entry, but sure enough there wasn't no chist. I had it all to sweep up next day.

' "He must have took it away to sea," says I to the cap'n, an' even then he didn't want to agree, but we was both beat out. I told him where I'd always seen Mis' Tolland get her money from, and we found much as a hundred dollars there in an old red morocco wallet. Cap'n John had been gone a good while a'ready, and she had spent what she needed. 'T was in an old desk o' his in the settin' room that we found the wallet.'

'At the last minute he may have taken his money to sea,' I suggested.

'Oh yes,' agreed Mrs Todd. 'He did take considerable to make his venture to bring home, as was customary, an' that was drowned with him as uncle agreed; but he had other property in shipping, and a thousand dollars invested in Portland in a cordage shop, but 't was about the time shipping begun to decay, and the cordage shop failed, and in the end I wa'n't so rich as I thought I was goin' to be for those few minutes on the cellar stairs. There was an auction that accumulated something. Old Mis' Tolland, the cap'n's mother, had heired some good furniture from a sister: there was above thirty chairs in all, and they're apt to sell well. I got over a thousand dollars when we come to settle up, and I made uncle take his five hundred; he was getting along in years and had met with losses in navigation, and he left it back to me when he died, so I had a real good lift. It all lays in the bank over to Rockland, and I draw my interest fall an' spring, with the little Mr Todd was able to leave me; but that's kind o' sacred money; 't was earnt and saved with the hope o' youth, an' I'm very particular what I spend it for. Oh yes, what with ownin' my house, I've been enabled to get along very well, with prudence!' said Mrs Todd contentedly.

'But there was the house and land,' I asked,—'what became of that part of the property?'

Mrs Todd looked into the fire, and a shadow of disapproval flitted over her face.

'Poor old uncle!' she said, 'he got childish about the matter. I was hoping to sell at first, and I had an offer, but he always run of an idea that there was more money hid away, and kept wanting me to delay; an' he used to go up there all alone and search, and dig in the cellar, empty an' bleak as 't was in winter weather or any time. An' he'd come and tell me he'd dreamed he found gold behind a stone in the cellar wall, or somethin'. And one night we all see the light o' fire up that way, an' the whole Landin' took the road, and run to look, and the Tolland property was all in a light blaze. I expect the old gentleman had dropped fire about; he said he'd been up there to see if everything was safe in the afternoon. As for the land, 't was so poor that everybody used to have a joke that the Tolland boys preferred to farm the sea instead. It's 'most all grown up to bushes now, where it ain't poor water grass in the low places. There's some upland that has a pretty view, after you cross the brook bridge. Years an' years after she died, there was some o' her flowers used to come up an' bloom in the door garden. I brought two or three that was unusual down here; they always come up and remind me of her, constant as the spring. But I never did want to fetch home that guitar, some way or 'nother; I wouldn't let it go at the auction, either. It was hangin' right there in the house when the fire took place. I've got some o' her other little things scattered about the house: that picture on the mantelpiece belonged to her.'

I had often wondered where such a picture had come from, and why Mrs Todd had chosen it; it was a French print of the statue of the Empress Josephine in the Savane at old Fort Royal, in Martinique.*

VI

MRS TODD drew her chair closer to mine; she held the cat and her knitting with one hand as she moved, but the cat was so warm and so sound asleep that she only stretched a lazy paw in spite of what must have felt like a slight earthquake. Mrs Todd began to speak almost in a whisper.

'I ain't told you all,' she continued; 'no, I haven't spoken of all to but very few. The way it came was this,' she said solemnly, and then stopped to listen to the wind, and sat for a moment in deferential silence, as if she waited for the wind to speak first. The cat suddenly lifted her head with quick excitement and gleaming eyes, and her mistress was leaning forward toward the fire with an arm laid on either knee, as if they were consulting the glowing coals for some augury. Mrs Todd looked like an old prophetess as she sat there with the firelight shining on her strong face; she was posed for some great painter. The woman with the cat was as unconscious and as mysterious as any sibyl of the Sistine Chapel.*

'There, that's the last struggle o' the gale,' said Mrs Todd, nodding her head with impressive certainty and still looking into the bright embers of the fire. 'You'll see!' She gave me another quick glance, and spoke in a low tone as if we might be overheard.

' 'T was such a gale as this the night Mis' Tolland died. She appeared more comfortable the first o' the evenin'; and Mrs Begg was more spent than I, bein' older, and a beautiful nurse that was the first to see and think of everything, but perfectly quiet an' never asked a useless question. You remember her funeral when you first come to the Landing? And she consented to goin' an' havin' a good sleep while she could, and left me one o' those good little pewter lamps that burnt whale oil an' made plenty o' light in the room, but not too bright to be disturbin'.

'Poor Mis' Tolland had been distressed the night before, an' all that day, but as night come on she grew more and more easy, an' was layin' there asleep; 't was like settin' by any sleepin' person, and I had none but usual thoughts. When the wind lulled and the rain, I could hear the seas, though more distant than this, and I don' know 's I observed any other sound than what the weather made; 't was a very solemn feelin' night. I set close by the bed; there was times she looked to find somebody when she was awake. The light was on her face, so I could see her plain; there was always times when she wore a look that made her seem a stranger you'd never set eyes on before. I did think what a world it was that her an' me should have come together so, and she have nobody but Dunnet Landin' folks about her in her extremity. "You're one o' the stray ones, poor creatur'," I said. I remember those very words passin' through my mind, but I saw reason to be glad she had some comforts, and didn't lack friends at the last, though she'd seen misery an' pain. I was glad she was quiet; all day she'd been restless, and we couldn't understand what she wanted from her French speech. We had the window open to give her air, an' now an' then a gust would strike that guitar that was on the wall and set it swinging by the blue ribbon, and soundin' as if somebody begun to play it. I come near takin' it down, but you never know what'll fret a sick person an' put 'em on the rack, an' that guitar was one o' the few things she'd brought with her.'

I nodded assent, and Mrs Todd spoke still lower.

'I set there close by the bed; I'd been through a good deal for some days back, and I thought I might 's well be droppin' asleep too, bein' a quick person to wake. She looked to me as if she might last a day longer, certain, now she'd got more comfortable, but I was real tired, an' sort o' cramped as watchers will get, an' a fretful feeling begun to creep over me such as they often do have. If you give way, there ain't no support for the sick person; they can't count on no composure o' their own. Mis' Tolland moved then, a little restless, an' I forgot me quick enough, an' begun to hum out a little part of a hymn tune just to make her feel everything was as usual an' not wake up into a poor uncertainty. All of a sudden she set right up in bed with her eyes wide open, an' I stood an' put my arm behind her; she hadn't moved like that for days. And she reached out both her arms toward the door, an' I looked the way she was lookin', an' I see some one was standin' there against the dark. No, 't wa'n't Mis' Begg; 't was somebody a good deal shorter than Mis' Begg. The lamplight struck across the room between us. I couldn't tell the shape, but 't was a woman's dark face lookin' right at us; 't wa'n't but an instant I could see. I felt dreadful cold, and my head begun to swim; I thought the light went out; 't wa'n't but an instant, as I say, an' when my sight come back I couldn't see nothing there. I was one that didn't know what it was to faint away, no matter what happened; time was I felt above it in others, but 't was somethin' that made poor human natur' quail. I saw very plain while I could see; 't was a pleasant enough face, shaped somethin' like Mis' Tolland's, and a kind of expectin' look.

'No, I don't expect I was asleep,' Mrs Todd assured me quietly, after a moment's pause, though I had not spoken. She gave a heavy sigh before she went on. I could see that the recollection moved her in the deepest way.

'I suppose if I hadn't been so spent an' quavery with long watchin', I might have kept my head an' observed much better,' she added humbly; 'but I see all I could bear. I did try to act calm, an' I laid Mis' Tolland down on her pillow, an' I was a-shakin' as I done it. All she did was to look up to me so satisfied and sort o' questioning, an' I looked back to her.

'"You saw her, didn't you?" she says to me, speakin' perfectly reasonable. "'T is my mother," she says again, very feeble, but lookin' straight up at me, kind of surprised with the pleasure, and smiling as if she saw I was overcome, an' would have said more if she could, but we had hold of hands. I see then her change was comin', but I didn't call Mis' Begg, nor make no uproar. I felt calm then, an' lifted to somethin' different as I never was since. She opened her eyes just as she was goin'—

' "You saw her, didn't you?" she said the second time, an' I says, *"Yes, dear, I did; you ain't never goin' to feel strange an' lonesome no more."* An' then in a few quiet minutes 't was all over. I felt they'd gone away together. No, I wa'n't alarmed afterward; 't was just that one moment I couldn't live under, but I never called it beyond reason I should see the other watcher. I saw plain enough there was somebody there with me in the room.

VII

' 'T WAS just such a night as this Mis' Tolland died,' repeated Mrs Todd, returning to her usual tone and leaning back comfortably in her chair as she took up her knitting. ' 'T was just such a night as this. I've told the circumstances to but very few; but I don't call it beyond reason. When folks is goin' 't is all natural, and only common things can jar upon the mind. You know plain enough there's somethin' beyond this world; the doors stand wide open. "There's somethin' of us that must still live on; we've got to join both worlds together an' live in one but for the other."* The doctor said that to me one day, an' I never could forget it; he said 't was in one o' his old doctor's books.'

We sat together in silence in the warm little room; the rain dropped heavily from the eaves, and the sea still roared, but the high wind had done blowing. We heard the far complaining fog horn of a steamer up the Bay.

'There goes the Boston boat out, pretty near on time,' said Mrs Todd with satisfaction. 'Sometimes these late August storms 'll sound a good deal worse than they really be. I do hate to hear the poor steamers callin' when they're bewildered in thick nights in winter, comin' on the coast. Yes, there goes the boat; they'll find it rough at sea, but the storm's all over.'

WILLIAM'S WEDDING

I

THE hurry of life in a large town, the constant putting aside of preference to yield to a most unsatisfactory activity, began to vex me, and one day I took the train, and only left it for the eastward-bound boat. Carlyle says somewhere that the only happiness a man ought to ask for is happiness enough to get his work done; and against this the complexity and futile ingenuity of social life seems a conspiracy.* But the first salt wind from the east, the first sight of a lighthouse set boldly on its outer rock, the flash of a gull, the waiting procession of seaward-bound firs on an island, made me feel solid and definite again, instead of a poor, incoherent being. Life was resumed, and anxious living blew away as if it had not been. I could not breathe deep enough or long enough. It was a return to happiness.

The coast had still a wintry look; it was far on in May, but all the shore looked cold and sterile. One was conscious of going north as well as east, and as the day went on the sea grew colder, and all the warmer air and bracing strength and stimulus of the autumn weather, and storage of the heat of summer, were quite gone. I was very cold and very tired when I came at evening up the lower bay, and saw the white houses of Dunnet Landing climbing the hill. They had a friendly look, these little houses, not as if they were climbing up the shore, but as if they were rather all coming down to meet a fond and weary traveler, and I could hardly wait with patience to step off the boat. It was not the usual eager company on the wharf. The coming-in of the mail-boat was the one large public event of a summer day, and I was disappointed at seeing none of my intimate friends but Johnny Bowden, who had evidently done nothing all winter but grow, so that his short sea-smitten clothes gave him a look of poverty.

Johnny's expression did not change as we greeted each other, but I suddenly felt that I had shown indifference and inconvenient delay by not coming sooner; before I could make an apology he took my small portmanteau, and walking before me in his old fashion he made straight up the hilly road toward Mrs Todd's. Yes, he was much grown—it had never occurred to me the summer before that Johnny was likely, with the help of time and other forces, to grow into a young man; he was such a well-framed and well-settled chunk of a boy that nature seemed to have set him aside as something finished, quite satisfactory and entirely completed.

The wonderful little green garden had been enchanted away by winter. There were a few frost-bitten twigs and some thin shrubbery against the fence, but it was a most unpromising small piece of ground. My heart was beating like a lover's as I passed it on the way to the door of Mrs Todd's house, which seemed to have become much smaller under the influence of winter weather.

'She hasn't gone away?' I asked Johnny Bowden with a sudden anxiety just as we reached the doorstep.

'Gone away!' he faced me with blank astonishment,—'I see her settin' by Mis' Caplin's window, the one nighest the road, about four o'clock!' And eager with suppressed news of my coming he made his entrance as if the house were a burrow.

Then on my homesick heart fell the voice of Mrs Todd. She stopped, through what I knew to be excess of feeling, to rebuke Johnny for bringing in so much mud, and I dallied without for one moment during the ceremony; then we met again face to face.

II

'I DARE say you can advise me what shapes they are going to wear. My meetin'-bunnit ain't going to do me again this year; no! I can't expect 't would do me forever,' said Mrs Todd, as soon as she could say anything. 'There! do set down and tell me how you have been! We've got a weddin' in the family, I s'pose you know?'

'A wedding!' said I, still full of excitement.

'Yes; I expect if the tide serves and the line storm* don't overtake him they'll come in and appear out on Sunday. I shouldn't have concerned me about the bunnit for a month yet, nobody would notice, but havin' an occasion like this I shall show consider'ble. 'T will be an ordeal for William!'

'For *William*!' I exclaimed. 'What do you mean, Mrs Todd?'

She gave a comfortable little laugh. 'Well, the Lord's seen reason at last an' removed Mis' Cap'n Hight up to the farm, an' I don't know but the weddin's going to be this week. Esther's had a great deal of business disposin' of her flock, but she's done extra well—the folks that owns the next place goin' up country are well off. 'T is elegant land north side o' that bleak ridge, an' one o' the boys has been Esther's right-hand man of late. She instructed him in all matters, and after she markets the early lambs he's goin' to take the farm on halves, an' she's give the refusal to him to buy her out within two years. She's reserved the buryin'-lot, an' the right o' way in, an'—'

I couldn't stop for details. I demanded reassurance of the central fact.

'William going to be married?' I repeated; whereat Mrs Todd gave me a searching look that was not without scorn.

'Old Mis' Hight's funeral was a week ago Wednesday, and 't was very well attended,' she assured me after a moment's pause.

'Poor thing!' said I, with a sudden vision of her helplessness and angry battle against the fate of illness; 'it was very hard for her.'

'I thought it was hard for Esther!' said Mrs Todd without sentiment.

III

I HAD an odd feeling of strangeness: I missed the garden, and the little rooms, to which I had added a few things of my own the summer before, seemed oddly unfamiliar. It was like the hermit crab in a cold new shell,—and with the windows shut against the raw May air, and a strange silence and grayness of the sea all that first night and day of my visit, I felt as if I had after all lost my hold of that quiet life.

Mrs Todd made the apt suggestion that city persons were prone to run themselves to death, and advised me to stay and get properly rested now that I had taken the trouble to come. She did not know how long I had been homesick for the conditions of life at the Landing the autumn before—it was natural enough to feel a little unsupported by compelling incidents on my return.

Some one has said that one never leaves a place, or arrives at one, until the next day! But on the second morning I woke with the familiar feeling of interest and ease, and the bright May sun was streaming in, while I could hear Mrs Todd's heavy footsteps pounding about in the other part of the house as if something were going to happen. There was the first golden robin singing somewhere close to the house, and a lovely aspect of spring now, and I looked at the garden to see that in the warm night some of its treasures had grown a hand's breadth; the determined spikes of yellow daffies stood tall against the doorsteps, and the bloodroot was unfolding leaf and flower. The belated spring which I had left behind farther south had overtaken me on this northern coast. I even saw a presumptuous dandelion in the garden border.

It is difficult to report the great events of New England; expression is so slight, and those few words which escape us in moments of deep feeling look but meagre on the printed page. One has to assume too much of the dramatic fervor as one reads; but as I came out of my room at breakfast-time I met Mrs Todd face to face, and when she said to me, 'This weather'll bring William in after her; 't is their happy day!' I felt something take possession of me which ought to communicate itself to the least sympathetic reader of this cold page. It is written for those who have a Dunnet Landing of their own: who either kindly share this with its writer, or possess another.

'I ain't seen his comin' sail yet; he'll be likely to dodge round among the islands so he'll be the less observed,' continued Mrs Todd. 'You can get a dory up the bay, even a clean new painted one, if you know as how, keepin' it against the high land.' She stepped to the door and looked off to sea as she spoke. I could see her eye follow the gray shores to and fro, and then a bright light spread over her calm face. 'There he comes, and he's striking right in across the open bay like a man!' she said with splendid approval. 'See, there he comes! Yes, there's William, and he's bent his new sail.'

I looked too, and saw the fleck of white no larger than a gull's wing yet, but present to her eager vision.

I was going to France for the whole long summer that year, and the more I thought of such an absence from these simple scenes the more dear and delightful they became. Santa Teresa says that the true proficiency of the soul is not in much thinking, but in much loving,* and sometimes I believed that I had never found love in its simplicity as I had found it at Dunnet Landing in the various hearts of Mrs Blackett and Mrs Todd and William. It is only because one came to know them, these three, loving and wise and true, in their own habitations. Their counterparts are in every village in the world, thank heaven, and the gift to one's life is only in its discernment. I had only lived in Dunnet until the usual distractions and artifices of the world were no longer in control, and I saw these simple natures clear. 'The happiness of life is in its recognitions. It seems that we are not ignorant of these truths, and even that we believe them; but we are so little accustomed to think of them, they are so strange to us—'*

'Well now, deary me!' said Mrs Todd, breaking into exclamation; 'I've got to fly round—I thought he'd have to beat; he can't sail far on that tack, and he won't be in for a good hour yet—I expect he's made every arrangement, but he said he shouldn't go up after Esther unless the weather was good, and I declare it did look doubtful this morning.'

I remembered Esther's weather-worn face. She was like a Frenchwoman who had spent her life in the fields. I remembered her pleasant look, her childlike eyes, and thought of the astonishment of joy she would feel now in being taken care of and tenderly sheltered from wind and weather after all these years. They were going to be young again now, she and

William, to forget work and care in the spring weather. I could hardly wait for the boat to come to land, I was so eager to see his happy face.

'Cake an' wine I'm goin' to set 'em out!' said Mrs Todd. 'They won't stop to set down for an ordered meal, they'll want to get right out home quick's they can. Yes, I'll give 'em some cake an' wine—I've got a rare plum-cake from my best receipt, and a bottle o' wine that the old Cap'n Denton of all give me, one of two, the day I was married, one we had and one we saved, and I've never touched it till now. He said there wa'n't none like it in the State o' Maine.'

It was a day of waiting, that day of spring; the May weather was as expectant as our fond hearts, and one could see the grass grow green hour by hour. The warm air was full of birds, there was a glow of light on the sea instead of the cold shining of chilly weather which had lingered late. There was a look on Mrs Todd's face which I saw once and could not meet again. She was in her highest mood. Then I went out early for a walk, and when I came back we sat in different rooms for the most part. There was such a thrill in the air that our only conversation was in her most abrupt and incisive manner. She was knitting, I believe, and as for me I dallied with a book. I heard her walking to and fro, and the door being wide open now, she went out and paced the front walk to the gate as if she walked a quarter-deck.

It is very solemn to sit waiting for the great events of life—most of us have done it again and again—to be expectant of life or expectant of death gives one the same feeling.

But at the last Mrs Todd came quickly back from the gate, and standing in the sunshine at the door, she beckoned me as if she were a sibyl.

'I thought you comprehended everything the day you was up there,' she added with a little more patience in her tone, but I felt that she thought I had lost instead of gained since we parted the autumn before.

'William's made this pretext o' goin' fishin' for the last time. 'T wouldn't done to take notice, 't would scared him to death! but there never was nobody took less comfort out o' forty years courtin'. No, he won't have to make no further pretexts,' said Mrs Todd, with an air of triumph.

'Did you know where he was going that day?' I asked with a sudden burst of admiration at such discernment.

'I did!' replied Mrs Todd grandly.

'Oh! but that pennyroyal lotion,' I indignantly protested, remembering that under pretext of mosquitoes she had be-smeared the poor lover in an awful way—why, it was outrageous! Medea could not have been more conscious of high ultimate purposes.*

'Darlin',' said Mrs Todd, in the excitement of my arrival and the great concerns of marriage, 'he's got a beautiful shaped face, and they pison him very unusual—you wouldn't have had him present himself to his lady all lop-sided with a mosquito-bite? Once when we was young I rode up with him, and they set upon him in concert the minute we entered the woods.' She stood before me reproachfully, and I was conscious of deserved rebuke. 'Yes, you've come just in the nick of time to advise me about a bunnit. They say large bows on top is liable to be worn.'

IV

THE period of waiting was one of direct contrast to these high moments of recognition. The very slowness of the morning hours wasted that sense of excitement with which we had begun the day. Mrs Todd came down from the mount where her face had shone so bright,* to the cares of common life, and some acquaintances from Black Island for whom she had little natural preference or liking came, bringing a poor, sickly child to get medical advice. They were noisy women with harsh, clamorous voices, and they stayed a long time. I heard the clink of teacups, however, and could detect no impatience in the tones of Mrs Todd's voice; but when they were at last going away, she did not linger unduly over her leave-taking, and returned to me to explain that they were people she had never liked, and they had made an excuse of a friendly visit to save their doctor's bill; but she pitied the poor little child, and knew beside that the doctor was away.

'I had to give 'em the remedies right out,' she told me; 'they wouldn't have bought a cent's worth o' drugs down to the store for that dwindlin' thing. She needed feedin' up, and I don't expect she gets milk enough; they're great butter-makers down to Black Island, 't is excellent pasturage, but they use no milk themselves, and their butter is heavy laden with salt to make weight, so that you'd think all their ideas come down from Sodom.'*

She was very indignant and very wistful about the pale little girl. 'I wish they'd let me kept her,' she said. 'I kind of advised it, and her eyes was so wishful in that pinched face when she heard me, so that I could see what was the matter with her, but they said she wa'n't prepared. Prepared!' And Mrs Todd snuffed like an offended war-horse, and departed; but I could hear her still grumbling and talking to herself in high dudgeon an hour afterward.

At the end of that time her arch enemy, Mari' Harris, appeared at the side-door with a gingham handkerchief over her head. She was always on hand for the news, and made some formal excuse for her presence,—she wished to borrow the weekly paper. Captain Littlepage, whose house-keeper she was, had taken it from the post-office in the morning, but had forgotten, being of failing memory, what he had done with it.

'How is the poor old gentleman?' asked Mrs Todd with solicitude, ignoring the present errand of Maria and all her concerns.

I had spoken the evening before of intended visits to Captain Littlepage and Elijah Tilley, and I now heard Mrs Todd repeating my inquiries and intentions, and fending off with unusual volubility of her own the curious questions that were sure to come. But at last Maria Harris secured an opportunity and boldly inquired if she had not seen William ashore early that morning.

'I don't say he wasn't,' replied Mrs Todd; 'Thu'sday's a very usual day with him to come ashore.'

'He was all dressed up,' insisted Maria—she really had no sense of propriety. 'I didn't know but they was going to be married?'

Mrs Todd did not reply. I recognized from the sounds that reached me that she had retired to the fastnesses of the kitchen-closet and was clattering the tins.

'I expect they'll marry soon anyway,' continued the visitor.

'I expect they will if they want to,' answered Mrs Todd. 'I don't know nothin' 't all about it; that's what folks say.' And presently the gingham handkerchief retreated past my window.

'I routed her, horse and foot,' said Mrs Todd proudly, coming at once to stand at my door. 'Who's coming now?' as two figures passed inward bound to the kitchen.

They were Mrs Begg and Johnny Bowden's mother, who were favorites, and were received with Mrs Todd's usual civilities. Then one of the Mrs Caplins came with a cup in hand to borrow yeast. On one pretext or another nearly all our acquaintance came to satisfy themselves of the facts, and see what Mrs Todd would impart about the wedding. But she firmly avoided the subject through the length of every call and errand, and answered the final leading question of each curious guest with her non-committal phrase, 'I don't know nothin' 't all about it; that's what folks say!'

She had just repeated this for the fourth or fifth time and shut the door upon the last comers, when we met in the little front entry. Mrs Todd was not in a bad temper, but highly amused. 'I've been havin' all sorts o' social privileges, you may have observed. They didn't seem to consider that if they could only hold out till afternoon they'd know as much as I did. There wa'n't but one o' the whole sixteen that showed real interest, the rest demeaned themselves to ask out o' cheap curiosity; no, there wa'n't but one showed any real feelin'.'

'Miss Maria Harris you mean?' and Mrs Todd laughed.

'Certain, dear,' she agreed, 'how you do understand poor human natur'!'

A short distance down the hilly street stood a narrow house that was newly painted white. It blinded one's eyes to catch the reflection of the sun. It was the house of the minister, and a wagon had just stopped before it; a man was helping a woman to alight, and they stood side by side for a moment, while Johnny Bowden appeared as if by magic, and climbed to the wagon-seat. Then they went into the house and shut the door. Mrs Todd and I stood close together and watched; the tears were running down her cheeks. I watched Johnny Bowden, who made light of so great a moment by so handling the whip that the old white Caplin horse started up from time to time and was inexorably stopped as if he had some idea of running away. There was something in the back of the wagon which now and then claimed the boy's attention; he leaned over as if there were something very precious left in his charge; perhaps it was only Esther's little trunk going to its new home.

At last the door of the parsonage opened, and two figures came out. The minister followed them and stood in the door-way, delaying them with parting words; he could not have thought it was a time for admonition.

'He's all alone; his wife's up to Portland to her sister's,' said Mrs Todd aloud, in a matter-of-fact voice. 'She's a nice woman, but she might ha' talked too much. There! see, they're comin' here. I didn't know how 't would be. Yes, they're comin' up to see us before they go home. I declare, if William ain't lookin' just like a king!'

Mrs Todd took one step forward, and we stood and waited. The happy pair came walking up the street, Johnny Bowden driving ahead. I heard a plaintive little cry from time to time to which in the excitement of the moment I had not stopped to listen; but when William and Esther had come and shaken hands with Mrs Todd and then with me, all in silence, Esther stepped quickly to the back of the wagon, and unfastening some cords returned to us carrying a little white lamb. She gave a shy glance at William as she fondled it and held it to her heart, and then, still silent, we went into the house together. The lamb had stopped bleating. It was lovely to see Esther carry it in her arms.

When we got into the house, all the repression of Mrs

Todd's usual manner was swept away by her flood of feeling. She took Esther's thin figure, lamb and all, to her heart and held her there, kissing her as she might have kissed a child, and then held out her hand to William and they gave each other the kiss of peace.* This was so moving, so tender, so free from their usual fetters of self-consciousness, that Esther and I could not help giving each other a happy glance of comprehension. I never saw a young bride half so touching in her happiness as Esther was that day of her wedding. We took the cake and wine of the marriage feast together, always in silence, like a true sacrament, and then to my astonishment I found that sympathy and public interest in so great an occasion were going to have their way. I shrank from the thought of William's possible sufferings, but he welcomed both the first group of neighbors and the last with heartiness; and when at last they had gone, for there were thoughtless loiterers in Dunnet Landing, I made ready with eager zeal and walked with William and Esther to the water-side. It was only a little way, and kind faces nodded reassuringly from the windows, while kind voices spoke from the doors. Esther carried the lamb on one arm; she had found time to tell me that its mother had died that morning and she could not bring herself to the thought of leaving it behind. She kept the other hand on William's arm until we reached the landing. Then he shook hands with me, and looked me full in the face to be sure I understood how happy he was, and stepping into the boat held out his arms to Esther—at last she was his own.

I watched him make a nest for the lamb out of an old seacloak at Esther's feet, and then he wrapped her own shawl round her shoulders, and finding a pin in the lapel of his Sunday coat he pinned it for her. She looked at him fondly while he did this, and then glanced up at us, a pretty, girlish color brightening her cheeks.

We stood there together and watched them go far out into the bay. The sunshine of the May day was low now, but there was a steady breeze, and the boat moved well.

'Mother'll be watching for them,' said Mrs Todd. 'Yes,

mother'll be watching all day, and waiting. She'll be so happy to have Esther come.'

We went home together up the hill, and Mrs Todd said nothing more; but we held each other's hand all the way.

OTHER FICTION

FAIR DAY

WIDOW Mercy Bascom came back alone into the empty kitchen and seated herself in her favorite splint-bottomed chair by the window, with a dreary look on her face.

'I s'pose I be an old woman, an' past goin' to cattle shows an' junketings, but folks needn't take it so for granted. I'm sure I don't want to be on my feet all day, trapesin' fair grounds an' swallowin' everybody's dust; not but what I'm as able as most, though I be seventy-three year old.'

She folded her hands in her lap and looked out across the deserted yard. There was not even a hen in sight; she was left alone for the day. 'Tobias's folks,' as she called the son's family with whom she made her home—Tobias's folks had just started for a day's pleasuring at the county fair, ten miles distant. She had not thought of going with them, nor expected any invitation; she had even helped them off with her famous energy; but there was an unexpected reluctance at being left behind, a sad little feeling that would rise suddenly in her throat as she stood in the door and saw them drive away in the shiny, two-seated wagon. Johnny, the youngest and favorite of her grandchildren, had shouted back in his piping voice, 'I wish you was goin', Grandma.'

'The only one on 'em that thought of me,' said Mercy Bascom to herself, and then not being a meditative person by nature, she went to work industriously and proceeded to the repairing of Tobias's work-day coat. It was sharp weather now in the early morning, and he would soon need the warmth of it. Tobias's placid wife never anticipated and always lived in a state of trying to catch up with her work. It never had been the elder woman's way, and Mercy reviewed her own active career with no mean pride. She had been left a widow at twenty-eight, with four children and a stony New Hampshire farm, but had bravely won her way, paid her debts, and provided the three girls and their brother Tobias with the best available schooling.

For a woman of such good judgment and high purpose in life, Mrs Bascom had made a very unwise choice in marrying Tobias Bascom the elder. He was not even the owner of a good name, and led her a terrible life with his drunken shiftlessness, and hindrance of all her own better aims. Even while the children were babies, however, and life was at its busiest and most demanding stages, the determined soul would not be baffled by such damaging partnership. She showed the plainer of what stuff she was made, and simply worked the harder and went her ways more fiercely. If it were sometimes whispered that she was unamiable, her wiser neighbors understood the power of will that was needed to cope with circumstances that would have crushed a weaker woman. As for her children, they were very fond of her in the undemonstrative New England fashion. Only the two eldest could remember their father at all, and after he was removed from this world Tobias Bascom left but slight proofs of having ever existed at all, except in the stern lines and premature aging of his wife's face.

The years that followed were years of hard work on the little farm, but diligence and perseverance had their reward. When the three daughters came to womanhood they were already skilled farmhouse keepers, and were dispatched for their own homes well equipped with feather-beds and home-spun linen and woolen. Mercy Bascom was glad to have them well settled, if the truth were known. She did not like to have her own will and law questioned or opposed, and when she sat down to supper alone with her son Tobias, after the last daughter's wedding, she had a glorious feeling of peace and satisfaction.

'There's a sight o' work left yet in the old ma'am,' she said to Tobias, in an unwontedly affectionate tone. 'I guess we shall keep house together as comfortable as most folks.' But Tobias grew very red in the face and bent over his plate.

'I don' know 's I want the girls to get ahead of me,' he said sheepishly. 'I ain't meanin' to put you out with another wedding right away, but I've been a-lookin' round, an' I guess I've found somebody to suit *me*.'

Mercy Bascom turned cold with misery and disappointment. 'Why T'bias,' she said, anxiously, 'folks always said that

you was cut out for an old bachelor till I come to believe it, an'
I've been lottin' on'—*

'Course nobody's goin' to wrench me an' you apart,' said
Tobias gallantly. 'I made up my mind long ago you an' me was
yoke-mates, mother. An' I had it in my mind to fetch you
somebody that would ease you o' quite so much work now
'Liza's gone off.'

'I don't want nobody,' said the grieved woman, and she
could eat no more supper; that festive supper for which she
had cooked her very best. Tobias was sorry for her, but he had
his rights, and now simply felt light-hearted because he had
freed his mind of this unwelcome declaration. Tobias was
slow and stolid to behold, but he was a man of sound ideas
and great talent for farming. He had found it difficult to
choose between his favorites among the marriageable girls, a
bright young creature who was really too good for him, but
penniless, and a weaker damsel who was heiress to the best
farm in town. The farm won the day at last; and Mrs Bascom
felt a thrill of pride at her son's worldly success; then she
asked to know her son's plans, and was wholly disappointed.
Tobias meant to sell the old place; he had no idea of leaving
her alone as she wistfully complained; he meant to have her
make a new home at the Bassett place with him and his bride.

That she would never do: the old place which had given
them a living never should be left or sold to strangers. Tobias
was not prepared for her fierce outburst of reproach at the
mere suggestion. She would live alone and pay her way as she
always had done, and so it was, for a few years of difficulties.
Tobias was never ready to plough or plant when she needed
him; his own great farm was more than he could serve properly. It grew more and more difficult to hire workmen, and
they were seldom worth their wages. At last Tobias's wife, who
was a kindly soul, persuaded her reluctant mother-in-law to
come and spend a winter; the old woman was tired and for
once disheartened; she found herself deeply in love with her
grandchildren, and so next spring she let the little hill farm
on the halves to an impecunious but hard-working young
couple.

To everybody's surprise the two women lived together

harmoniously. Tobias's wife did everything to please her mother-in-law except to be other than a Bassett. And Mercy, for the most part, ignored this misfortune, and rarely was provoked into calling it a fault. Now that the necessity for hard work and anxiety was past, she appeared to have come to an Indian summer shining-out of her natural amiability and tolerance. She was sometimes indirectly reproachful of her daughter's easy-going ways, and set an indignant example now and then by a famous onslaught of unnecessary work, and always dressed and behaved herself in plainest farm fashion, while Mrs Tobias was given to undue worldliness and style. But they worked well together in the main, for, to use Mercy's own works, she 'had seen enough of life not to want to go into other folks' houses and make trouble.'

As people grow older their interests are apt to become fewer, and one of the thoughts that came oftenest to Mercy Bascom in her old age was a time-honored quarrel with one of her husband's sisters, who had been her neighbor many years before, and then moved to greater prosperity at the other side of the county. It is not worth while to tell the long story of accusations and misunderstandings, but while the two women did not meet for almost half a lifetime the grievance was as fresh as if it were yesterday's. Wrongs of defrauded sums of money and contested rights in unproductive acres of land, wrongs of slighting remarks and contempt of equal claims; the remembrance of all these was treasured as a miser fingers his gold. Mercy Bascom freed herself from the wearisome detail of every-day life whenever she could find a patient listener to whom to tell the long story. She found it as interesting as a story of the Arabian Nights, or an exciting play at the theatre. She would have you believe that she was faultless in the matter, and would not acknowledge that her sister-in-law Ruth Bascom, now Mrs Parlet, was also a hard-working woman with dependent little children at the time of the great fray.

Of late years her son had suspected that his mother regretted the alienation, but he knew better than to suggest a peace-making. 'Let them work—let them work!' he told his wife, when she proposed one night to bring the warring sisters-in-law unexpectedly together. It may have been that old Mercy began to feel a little lonely and would be glad to have somebody of her own age with whom to talk over old times. She never had known the people much in this Bassett region, and there were few but young folks left at any rate.

As the pleasure-makers hastened toward the fair that bright October morning Mercy sat by the table sewing at a sufficient patch in the old coat. There was little else to do all day but to get herself a luncheon at noon and have supper ready when the family came home cold and tired at night. The two cats came purring about her chair; one persuaded her to open the cellar door, and the other leaped to the top of the kitchen table unrebuked, and blinked herself to sleep there in the sun. This was a favored kitten brought from the old home, and seemed like a link between the old days and these. Her mistress noticed with surprise that pussy was beginning to look old, and she could not resist a little sigh. 'Land! the next world may seem dreadful new too, and I've got to get used to that,' she thought with a grim smile of foreboding. 'How do folks live that wants always to be on the go? There was Ruth Parlet, that must be always a visitin' and goin'—well, I won't say that there wasn't a time when I wished for the chance.' Justice always won the day in such minor questions as this.

Ruth Parlet's name started the usual thoughts, but somehow or other Mercy could not find it in her heart to be as harsh as usual. She remembered one thing after another about their girlhood together. They had been great friends then, and the animosity may have had its root in the fact that Ruth helped forward her brother's marriage. But there were years before that of friendly foregathering and girlish alliances and rivalries; spinning and herb gathering and quilting. It seemed, as Mercy thought about it, that Ruth was good company after all. But what did make her act so, and turn right round later on?

The morning grew warm, and at last Mrs Bascom had to open the window to let out the buzzing flies and an imprisoned wild bee. The patch was finished and the elbow would serve Tobias as good as new. She laid the coat over a chair and put her bent brass thimble into the paper-collar

box that served as work-basket. She used to have a queer
splint basket at the old place, but it had been broken under
something heavier when her household goods were moved.
Some of the family had long been tired of hearing that basket
regretted, and another had never been found worthy to take
its place. The thimble, the smooth mill bobbin on which was
wound black linen thread, the dingy lump of beeswax,* and a
smart leather needle-book, which Johnny had given her the
Christmas before, all looked ready for use, but Mrs Bascom
pushed them farther back on the table and quickly rose to her
feet. ''T ain't nine o'clock yet,' she said, exultantly. 'I'll just
take a couple o' crackers in my pocket and step over to the old
place. I'll take my time and be back soon enough to make 'em
that pan o' my hot gingerbread they'll be counting on for
supper.'

Half an hour later one might have seen a bent figure lock
the side door of the large farmhouse carefully, trying the
latch again and again to see if it were fast, putting the key into
a safe hiding-place by the door, and then stepping away up
the road with eager determination. 'I ain't felt so like a jaunt
this five year,' said Mercy to herself, 'an' if Tobias was here an'
Ann, they'd take all the fun out fussin' and talkin', an' bein'
afeard I'd tire myself, or wantin' me to ride over. I do like to
be my own master once in a while.'

The autumn day was glorious, with a fine flavor of fruit and
ripeness in the air. The sun was warm, there was a cool breeze
from the great hills, and far off across the wide valley the old
woman could see her little gray house on its pleasant eastern
slope; she could even trace the outline of the two small fields
and large pasture. 'I done well with it, if I wasn't nothin' but
a woman with four dependin' on me an' no means,' said
Mercy proudly as she came in full sight of the old place. It was
a long drive from one farm to the other by roundabout
highways, but there was a footpath known to the wayfarer
which took a good piece off the distance. 'Now, ain't this a
sight better than them hustlin' fairs?' Mercy asked gleefully as
she felt herself free and alone in the wide meadow-land. She
had long been promising little Johnny to take him over to
Gran'ma's house, as she loved to call it still. She could not

help thinking longingly how much he would enjoy this escapade. 'Why, I'm running away just like a young-one, that's what I be,' she exclaimed, and then laughed aloud for very pleasure.

The weather-beaten farmhouse was deserted that day, as its former owner suspected. She boldly gathered some of her valued spice-apples, with an assuring sense of proprietorship as she crossed the last narrow field. The Browns, man and wife and little boy and baby, had hied them early to the fair with nearly the whole population of the countryside. The house and yard and out-buildings never had worn such an aspect of appealing pleasantness as when Mercy Bascom came near. She felt as if she were going to cry for a minute, and then hurried to get inside the gate. She saw the outgoing track of horses' feet with delight, but went discreetly to the door and knocked, to make herself perfectly sure that there was no one left at home. Out of breath and tired as she was, she turned to look off at the view. Yes, there was Tobias's place, prosperous and white-painted; she could just get a glimpse of the upper roofs and gables. It was always a sorrow and complaint that a low hill kept her from looking up at this farm from any of the windows, but now that she was at the farm itself she found herself regarding Tobias's home with a good deal of affection. She looked sharply with an apprehension of fire, but there was no whiff of alarming smoke against the clear sky.

'Now I must git me a drink o' that water first of anything,' and she hastened to the creaking well-sweep and lowered the bucket. There was the same rusty, handleless tin dipper that she had left years before, standing on the shelf inside the well-curb. She was proud to find that the bucket was no heavier than ever, and was heartily thankful for the clear water. There never was such a well as that, and it seemed as if she had not been away a day. 'What an old gal I be,' said Mercy, with plaintive merriment. 'Well, they ain't made no great changes since I was here last spring,' and then she went over and held her face close against one of the kitchen windows, and took a hungry look at the familiar room. The bedroom door was open and a new sense of attachment to the place filled her heart. 'It seems as if I was locked out o' my own home,' she whispered as she looked in.

There were the same old spruce and pine boards that she had scrubbed so many times and trodden thin as she hurried to and fro about her work. It was very strange to see an unfamiliar chair or two, but the furnishings of a farm kitchen were much the same, and there was no great change. Even the cradle was like that cradle in which her own children had been rocked. She gazed and gazed, poor old Mother Bascom, and forgot the present as her early life came back in vivid memories. At last she turned away from the window with a sigh.

The flowers that she had planted herself long ago had bloomed all summer in the garden; there were still some ragged sailors and the snowberries and phlox and her favorite white mallows, of which she picked herself a posy. 'I'm glad the old place is so well took care of,' she thought, gratefully. 'An' they've new-silled the old barn I do declare, and battened the cracks to keep the dumb creatures warm. 'T was a sham-built barn anyways, but 't was the best I could do when the child'n needed something every handturn o' the day. It put me to some expense every year, tinkering of it up where the poor lumber warped and split. There, I enjoyed try'n to cope with things and gettin' the better of my disadvantages! The ground's too rich for me over there to Tobias's; I don't want things too easy, for my part. I feel most as young as ever I did, and I ain't a-goin' to play helpless, not for nobody.

'I declare for 't, I mean to come up here by an' by a spell an' stop with the young folks, an' give 'em a good lift with their work. I ain't needed all the time to Tobias's now, and they can hire help, while these can't. I've been favoring myself till I'm as soft as an old hoss that's right out of pasture an' can't pull two wheels without wheezin'.'

There was a sense of companionship in the very weather. The bees were abroad as if it were summer, and a flock of little birds came fluttering down close to Mrs Bascom as she sat on the doorstep. She remembered the biscuits in her pocket and ate them with a hunger she had seldom known of late, but she threw the crumbs generously to her feathered neighbors. The soft air, the brilliant or fading colors of the wide landscape, the comfortable feeling of relationship to her surroundings all served to put good old Mercy into a most peaceful state. There was only one thought that would not let her be quite happy. She could not get her sister-in-law Ruth Parlet out of her mind. And strangely enough the old grudge did not present itself with the usual power of aggravation; it was of their early friendship and Ruth's good fellowship that memories would come.

'I declare for 't, I wouldn't own up to the folks, but I should like to have a good visit with Ruth if so be that we could set aside the past,' she said, resolutely at last. 'I never thought I should come to it, but if she offered to make peace I wouldn't do nothin' to hinder it. Not to say but what I should have to free my mind on one or two points before we could start fair. I've waited forty year to make one remark to Ruthy Parlet. But there! we're gettin' to be old folks.' Mercy rebuked herself gravely. 'I don't want to go off with hard feelin's to nobody.' Whether this was the culmination of a long, slow process of reconciliation, or whether Mrs Bascom's placid satisfaction helped to hasten it by many stages, nobody could say. As she sat there she thought of many things; her life spread itself out like a picture; perhaps never before had she been able to detach herself from her immediate occupation in this way. She never had been aware of her own character and exploits to such a degree, and the minutes sped by as she thought with deep interest along the course of her own history. There was nothing she was ashamed of to an uncomfortable degree but the long animosity between herself and the children's aunt. How harsh she had been sometimes; she had even tried to prejudice everybody who listened to these tales of an offender. 'I wa'n't more 'n half right, now I come to look myself full in the face,' said Mercy Bascom, 'and I never owned it till this day.'

The sun was already past noon, and the good woman dutifully rose and with instant consciousness of resource glanced in at the kitchen window to tell the time by a familiar mark on the floor. 'I needn't start just yet,' she muttered. 'Oh my! how

I do wish I could git in and poke round into every corner! 'T would make this day just perfect.'

'There now!' she continued, 'p'raps they leave the key just where our folks used to.' And in another minute the key lay in Mercy's worn old hand. She gave a shrewd look along the road, opened the door, which creaked what may have been a hearty welcome, and stood inside the dear old kitchen. She had not been in the house alone since she left it, but now she was nobody's guest. It was like some shell-fish finding its own old shell again and settling comfortably into the convolutions. Even we must not follow Mother Bascom about from the dark cellar to the hot little attic. She was not curious about the Browns' worldly goods; indeed, she was nearly unconscious of anything but the comfort of going up and down the short flight of stairs and looking out of her own windows with nobody to watch.

'There's the place where Tobias scratched the cupboard door with a nail. Didn't I thrash him for it good?' she said once with a proud remembrance of the time when she was a lawgiver and proprietor and he dependent.

At length a creeping fear stole over her lest the family might return. She stopped one moment to look back into the little bedroom. 'How good I did use to sleep here,' she said. 'I worked as stout as I could the day through, and there wa'n't no wakin' up by two o'clock in the morning, and smellin' for fire and harkin' for thieves like I have to nowadays.'

Mercy stepped away down the long sloping field like a young woman. It was a long walk back to Tobias's, even if one followed the pleasant footpaths across country. She was heavy-footed, but entirely light-hearted when she came safely in at the gate of the Bassett place. 'I've done extra for me,' she said as she put away her old shawl and bonnet; 'but I'm goin' to git the best supper Tobias's folks have eat for a year,' and so she did.

'I've be'n over to the old place to-day,' she announced bravely to her son, who had finished his work and his supper and was now tipped back in his wooden arm-chair against the wall.

'You ain't, mother!' responded Tobias, with instant excitement. 'Next fall, then, I won't take no for an answer but what you'll go to the fair and see what's goin'. You ain't footed it way over there?'

Mother Bascom nodded. 'I have,' she answered solemnly, a minute later, as if the nod were not enough. 'T'bias, son,' she added, lowering her voice, 'I ain't one to give in my rights, but I was thinkin' it all over about y'r Aunt Ruth Parlet'—

'Now if that ain't curi's!' exclaimed Tobias, bringing his chair down hastily upon all four legs. 'I didn't know just how you'd take it, mother, but I see Aunt Ruth to-day to the fair, and she made everything o' me and wanted to know how you was, and she got me off from the rest, an' says she: "I declare I should like to see your marm again. I wonder if she won't agree to let bygones be bygones."'

'My sakes!' said Mercy, who was startled by this news. ''T is the hand o' Providence! How did she look, son?'

'A sight older 'n you look, but kind of natural too. One o' her sons' wives that she's made her home with, has led her a dance, folks say.'

'Poor old creatur'! we'll have her over here, if your folks don't find fault. I've had her in my mind'—

Tobias's folks, in the shape of his wife and little Johnny, appeared from the outer kitchen. 'I haven't had such a supper I don't know when,' repeated the younger woman for at least the fifth time. 'You must have been keepin' busy all day, Mother Bascom.'

But Mother Bascom and Tobias looked at each other and laughed.

'I ain't had such a good time I don't know when, but my feet are all of a fidget now, and I've got to git to bed. I've be'n runnin' away since you've be'n gone, Ann!' said the pleased old soul, and then went away, still laughing, to her own room. She was strangely excited and satisfied, as if she had at last paid a long-standing debt. She could trudge across pastures as well as anybody, and the old grudge was done with. Mercy hardly noticed how her fingers trembled as she unhooked the old gray gown. The odor of sweet fern shook out fresh and strong as she smoothed and laid it carefully over a chair.

There was a little rent in the skirt, but she could mend it by daylight.

The great harvest moon was shining high in the sky, and she needed no other light in the bedroom. 'I've be'n a smart woman to work in my day, and I've airnt a little pleasurin',' said Mother Bascom sleepily to herself. 'Poor Ruthy! so she looks old, does she? I'm goin' to tell her right out, 't was I that spoke first to Tobias.'

THE FLIGHT OF BETSEY LANE

I

ONE windy morning in May, three old women sat together near an open window in the shed chamber of Byfleet Poorhouse. The wind was from the northwest, but their window faced the southeast, and they were only visited by an occasional pleasant waft of fresh air. They were close together, knee to knee, picking over a bushel of beans, and commanding a view of the dandelion-starred, green yard below, and of the winding, sandy road that led to the village, two miles away. Some captive bees were scolding among the cobwebs of the rafters overhead, or thumping against the upper panes of glass; two calves were bawling from the barnyard, where some of the men were at work loading a dump-cart and shouting as if every one were deaf. There was a cheerful feeling of activity, and even an air of comfort, about the Byfleet Poor-house. Almost every one was possessed of a most interesting past, though there was less to be said about the future. The inmates were by no means distressed or unhappy; many of them retired to this shelter only for the winter season, and would go out presently, some to begin such work as they could still do, others to live in their own small houses; old age had impoverished most of them by limiting their power of endurance; but far from lamenting the fact that they were town charges, they rather liked the change and excitement of a winter residence on the poor-farm. There was a sharp-faced, hard-worked young widow with seven children, who was an exception to the general level of society, because she deplored the change in her fortunes. The older women regarded her with suspicion, and were apt to talk about her in moments like this, when they happened to sit together at their work.

The three bean-pickers were dressed alike in stout brown ginghams, checked by a white line, and all wore great faded aprons of blue drilling, with sufficient pockets convenient to the right hand. Miss Peggy Bond was a very small, belligerent-looking person, who wore a huge pair of steel-bowed spectacles, holding her sharp chin well up in air, as if no supplement an inadequate nose. She was more than half blind, but the spectacles seemed to face upward instead of square ahead, as if their wearer were always on the sharp lookout for birds. Miss Bond had suffered much personal damage from time to time, because she never took heed where she planted her feet, and so was always tripping and stubbing her bruised way through the world. She had fallen down hatchways and cellarways, and stepped composedly into deep ditches and pasture brooks; but she was proud of stating that she was upsighted, and so was her father before her. At the poorhouse, where an unusual malady was considered a distinction, upsightedness was looked upon as a most honorable infirmity. Plain rheumatism, such as afflicted Aunt Lavina Dow, whose twisted hands found even this light work difficult and tiresome,—plain rheumatism was something of every-day occurrence, and nobody cared to hear about it. Poor Peggy was a meek and friendly soul, who never put herself forward; she was just like other folks, as she always loved to say, but Mrs Lavina Dow was a different sort of person altogether, of great dignity and, occasionally, almost aggressive behavior. The time had been when she could do a good day's work with anybody: but for many years now she had not left the town-farm, being too badly crippled to work; she had no relations or friends to visit, but from an innate love of authority she could not submit to being one of those who are forgotten by the world. Mrs Dow was the hostess and social lawgiver here, where she remembered every inmate and every item of interest for nearly forty years, besides an immense amount of town history and biography for three or four generations back.

She was the dear friend of the third woman, Betsey Lane; together they led thought and opinion—chiefly opinion— and held sway, not only over Byfleet Poor-farm, but also the selectmen and all others in authority. Betsey Lane had spent most of her life as aid-in-general to the respected household of old General Thornton. She had been much trusted and valued, and, at the breaking up of that once large and flourishing family, she had been left in good circumstances, what with legacies and her own comfortable savings; but by sad misfortune and lavish generosity everything had been scattered, and after much illness, which ended in a stiffened arm and more uncertainty, the good soul had sensibly decided that it was easier for the whole town to support her than for a part of it. She had always hoped to see something of the world before she died; she came of an adventurous, seafaring stock, but had never made a longer journey than to the towns of Danby and Northville, thirty miles away.

They were all old women; but Betsey Lane, who was sixtynine, and looked much older, was the youngest. Peggy Bond was far on in the seventies, and Mrs Dow was at least ten years older. She made a great secret of her years; and as she sometimes spoke of events prior to the Revolution with the assertion of having been an eye-witness, she naturally wore an air of vast antiquity. Her tales were an inexpressible delight to Betsey Lane, who felt younger by twenty years because her friend and comrade was so unconscious of chronological limitations.

The bushel basket of cranberry beans was within easy reach, and each of the pickers had filled her lap from it again and again. The shed chamber was not an unpleasant place in which to sit at work, with its traces of seed corn hanging from the brown cross-beams, its spare churns, and dusty loom, and rickety wool-wheels, and a few bits of old furniture. In one far corner was a wide board of dismal use and suggestion, and close beside it an old cradle. There was a battered chest of drawers where the keeper of the poor-house kept his garden-seeds, with the withered remains of three seed cucumbers ornamenting the top. Nothing beautiful could be discovered, nothing interesting, but there was something usable and homely about the place. It was the favorite and untroubled bower of the bean-pickers, to which they might retreat unmolested from the public apartments of this rustic institution.

Betsey Lane blew away the chaff from her handful of beans. The spring breeze blew the chaff back again, and sifted it over her face and shoulders. She rubbed it out of her eyes impatiently, and happened to notice old Peggy holding her own handful high, as if it were an oblation, and turning her queer, up-tilted head this way and that, to look at the beans sharply, as if she were first cousin to a hen.

'There, Miss Bond, 't is kind of botherin' work for you, ain't it?' Betsey inquired compassionately.

'I feel to enjoy it, anything that I can do my own way so,' responded Peggy. 'I like to do my part. Ain't that old Mis' Fales comin' up the road? It sounds like her step.'

The others looked, but they were not far-sighted, and for a moment Peggy had the advantage. Mrs Fales was not a favorite.

'I hope she ain't comin' here to put up this spring. I guess she won't now, it's gettin' so late,' said Betsey Lane. 'She likes to go rovin' soon as the roads is settled.'

' 'T is Mis' Fales!' said Peggy Bond, listening with solemn anxiety. 'There, do let's pray her by!'

'I guess she's headin' for her cousin's folks up Beech Hill way,' said Betsey presently. 'If she'd left her daughter's this mornin', she'd have got just about as far as this. I kind o' wish she had stepped in just to pass the time o' day, long 's she wa'n't going to make no stop.'

There was a silence as to further speech in the shed chamber; and even the calves were quiet in the barnyard. The men had all gone away to the field where corn-planting was going on. The beans clicked steadily into the wooden measure at the pickers' feet. Betsey Lane began to sing a hymn, and the others joined in as best they might, like autumnal crickets; their voices were sharp and cracked, with now and then a few low notes of plaintive tone. Betsey herself could sing pretty well, but the others could only make a kind of accompaniment. Their voices ceased altogether at the higher notes.

'Oh my! I wish I had the means to go to the Centennial,'* mourned Betsey Lane, stopping so suddenly that the others had to go on croaking and shrilling without her for a moment before they could stop. 'It seems to me as if I can't die happy 'less I do,' she added; 'I ain't never seen nothin' of the world, an' here I be.'

'What if you was as old as I be?' suggested Mrs Dow pompously. 'You've got time enough yet, Betsey; don't you go an' despair. I knowed of a woman that went clean round the world four times when she was past eighty, an' enjoyed herself real well. Her folks followed the sea; she had three sons an' a daughter married,—all shipmasters, and she'd been with her own husband when they was young. She was left a widder early, and fetched up her family herself,—a real stirrin', smart woman. After they'd got married off, an' settled, an' was doing well, she come to be lonesome; and first she tried to stick it out alone, but she wa'n't one that could; an' she got a notion she hadn't nothin' before her but her last sickness, and she wa'n't a person that enjoyed havin' other folks do for her. So one on her boys—I guess 't was the oldest—said he was going to take her to sea; there was ample room, an' he was sailin' a good time o' year for the Cape o' Good Hope an' way up to some o' them tea-ports in the Chiny Seas. She was all high to go, but it made a sight o' talk at her age; an' the minister made it a subject o' prayer the last Sunday, and all the folks took a last leave; but she said to some she'd fetch 'em home something real pritty, and so did. An' then they come home t' other way, round the Horn, an' she done so well, an' was such a sight o' company, the other child'n was jealous, an' she promised she'd go a v'y'ge long o' each on 'em. She was as sprightly a person as ever I see; an' could speak well o' what she'd seen.'

'Did she die to sea?' asked Peggy, with interest.

'No, she died to home between v'y'ges, or she'd gone to sea again. I was to her funeral. She liked her son George's ship the best; 't was the one she was going on to Callao.* They said the men aboard all called her "gran'ma'am," an' she kep' 'em mended up, an' would go below and tend to 'em if they was sick. She might 'a' been alive an' enjoyin' of herself a good many years but for the kick of a cow; 't was a new cow out of a drove, a dreadful unruly beast.'

Mrs Dow stopped for breath, and reached down for a new supply of beans; her empty apron was gray with soft chaff. Betsey Lane, still pondering on the Centennial, began to sing another verse of her hymn, and again the old women joined her. At this moment some strangers came driving round into the yard from the front of the house. The turf was soft, and our friends did not hear the horses' steps. Their voices cracked and quavered; it was a funny little concert, and a lady in an open carriage just below listened with sympathy and amusement.

II

'BETSEY! Betsey! Miss Lane!' a voice called eagerly at the foot of the stairs that led up from the shed. 'Betsey! There's a lady here wants to see you right away.'

Betsey was dazed with excitement, like a country child who knows the rare pleasure of being called out of school. 'Lor', I ain't fit to go down, be I?' she faltered, looking anxiously at her friends; but Peggy was gazing even nearer to the zenith than usual, in her excited effort to see down into the yard, and Mrs Dow only nodded somewhat jealously, and said that she guessed 't was nobody would do her any harm. She rose ponderously, while Betsey hesitated, being, as they would have said, all of a twitter. 'It is a lady, certain,' Mrs Dow assured her; ''t ain't often there's a lady comes here.'

'While there was any of Mis' Gen'ral Thornton's folks left, I wa'n't without visits from the gentry,' said Betsey Lane, turning back proudly at the head of the stairs, with a touch of old-world pride and sense of high station. Then she disappeared, and closed the door behind her at the stair-foot with a decision quite unwelcome to the friends above.

'She needn't 'a' been so dreadful 'fraid anybody was goin' to listen. I guess we've got folks to ride an' see us, or had once, if we hain't now,' said Miss Peggy Bond, plaintively.

'I expect 't was only the wind shoved it to,' said Aunt Lavina. 'Betsey is one that gits flustered easier than some. I wish 't was somebody to take her off an' give her a kind of a good time; she's young to settle down 'long of old folks like us. Betsey's got a notion o' rovin' such as ain't my natur', but I should like to see her satisfied. She'd been a very understandin' person, if she had the advantages that some does.'

''T is so,' said Peggy Bond, tilting her chin high. 'I suppose you can't hear nothin' they're saying? I feel my hearin' ain't up to whar it was. I can hear things close to me well as ever;

but there, hearin' ain't everything; 't ain't as if we lived where there was more goin' on to hear. Seems to me them folks is stoppin' a good while.'

'They surely be,' agreed Lavina Dow.

'I expect it's somethin' particular. There ain't none of the Thornton folks left, except one o' the gran'darters, an' I've often heard Betsey remark that she should never see her more, for she lives to London. Strange how folks feels contented in them strayaway places off to the ends of the airth.'

The flies and bees were buzzing against the hot window-panes; the handfuls of beans were clicking into the brown wooden measure. A bird came and perched on the window-sill, and then flitted away toward the blue sky. Below, in the yard, Betsey Lane stood talking with the lady. She had put her blue drilling apron over her head, and her face was shining with delight.

'Lor', dear,' she said, for at least the third time, 'I remember ye when I first see ye; an awful pritty baby you was, an' they all said you looked just like the old gen'ral. Be you goin' back to foreign parts right away?'

'Yes, I'm going back; you know that all my children are there. I wish I could take you with me for a visit,' said the charming young guest. 'I'm going to carry over some of the pictures and furniture from the old house; I didn't care half so much for them when I was younger as I do now. Perhaps next summer we shall all come over for a while. I should like to see my girls and boys playing under the pines.'

'I wish you re'lly was livin' to the old place,' said Betsey Lane. Her imagination was not swift; she needed time to think over all that was being told her, and she could not fancy the two strange houses across the sea. The old Thornton house was to her mind the most delightful and elegant in the world.

'Is there anything I can do for you?' asked Mrs Strafford kindly,—'anything that I can do for you myself, before I go away? I shall be writing to you, and sending some pictures of the children, and you must let me know how you are getting on.'

'Yes, there is one thing, darlin'. If you could stop in the village an' pick me out a pritty, little, small lookin'-glass, that I can keep for my own an' have to remember you by. 'T ain't that I want to set me above the rest o' the folks, but I was always used to havin' my own when I was to your grandma's. There's very nice folks here, some on 'em, and I'm better off than if I was able to keep house; but sence you ask me, that's the only thing I feel cropin' about.* What be you goin' right back for? ain't you goin' to see the great fair to Pheladelphy, that everybody talks about?'

'No,' said Mrs Strafford, laughing at this eager and almost convicting question. 'No; I'm going back next week. If I were, I believe that I should take you with me. Good-by, dear old Betsey; you make me feel as if I were a little girl again; you look just the same.'

For full five minutes the old woman stood out in the sunshine, dazed with delight, and majestic with a sense of her own consequence. She held something tight in her hand, without thinking what it might be; but just as the friendly mistress of the poor-farm came out to hear the news, she tucked the roll of money into the bosom of her brown gingham dress. ' 'T was my dear Mis' Katy Strafford,' she turned to say proudly. 'She come way over from London; she's been sick; they thought the voyage would do her good. She said most the first thing she had on her mind was to come an' find me, and see how I was, an' if I was comfortable; an' now she's goin' right back. She's got two splendid houses; an' said how she wished I was there to look after things,—she remembered I was always her gran'ma's right hand. Oh, it does so carry me back, to see her! Seems if all the rest on 'em must be there together to the old house. There, I must go right up an' tell Mis' Dow an' Peggy.'

'Dinner's all ready; I was just goin' to blow the horn for the men-folks,' said the keeper's wife. 'They'll be right down. I expect you've got along smart with them beans,—all three of you together;' but Betsey's mind roved so high and so far at that moment that no achievements of bean-picking could lure it back.

III

THE long table in the great kitchen soon gathered its company of waifs and strays,—creatures of improvidence and misfortune, and the irreparable victims of old age. The dinner was satisfactory, and there was not much delay for conversation. Peggy Bond and Mrs Dow and Betsey Lane always sat together at one end, with an air of putting the rest of the company below the salt.* Betsey was still flushed with excitement; in fact, she could not eat as much as usual, and she looked up from time to time expectantly, as if she were likely to be asked to speak of her guest; but everybody was hungry, and even Mrs Dow broke in upon some attempted confidences by asking inopportunely for a second potato. There were nearly twenty at the table, counting the keeper and his wife and two children, noisy little persons who had come from school with the small flock belonging to the poor widow, who sat just opposite our friends. She finished her dinner before any one else, and pushed her chair back; she always helped with the housework,—a thin, sorry, bad-tempered-looking poor soul, whom grief had sharpened instead of softening. 'I expect you feel too fine to set with common folks,' she said enviously to Betsey.

'Here I be a-settin',' responded Betsey calmly. 'I don' know 's I behave more unbecomin' than usual.' Betsey prided herself upon her good and proper manners; but the rest of the company, who would have liked to hear the bit of morning news, were now defrauded of that pleasure. The wrong note had been struck; there was a silence after the clatter of knives and plates, and one by one the cheerful town charges disappeared. The bean-picking had been finished, and there was a call for any of the women who felt like planting corn; so Peggy Bond, who could follow the line of hills pretty fairly, and Betsey herself, who was still equal to anybody at that work, and Mrs Dow, all went out to the field together. Aunt Lavina labored slowly up the yard, carrying a light splint-bottomed kitchen chair and her knitting-work, and sat near the stone wall on a gentle rise, where she could see the pond and the green country, and exchange a word with her friends as they came and went up and down the rows. Betsey vouchsafed a word now and then about Mrs Strafford, but you would have thought that she had been suddenly elevated to Mrs Strafford's own cares and the responsibilities attending them, and had little in common with her old associates. Mrs Dow and Peggy knew well that these high-feeling times never lasted long, and so they waited with as much patience as they could muster. They were by no means without that true tact which is only another word for unselfish sympathy.

The strip of corn land ran along the side of a great field; at the upper end of it was a field-corner thicket of young maples and walnut saplings, the children of a great nut-tree that marked the boundary. Once, when Betsey Lane found herself alone near this shelter at the end of her row, the other planters having lagged behind beyond the rising ground, she looked stealthily about, and then put her hand inside her gown, and for the first time took out the money that Mrs Strafford had given her. She turned it over and over with an astonished look: there were new bank-bills for a hundred dollars. Betsey gave a funny little shrug of her shoulders, came out of the bushes, and took a step or two on the narrow edge of turf, as if she were going to dance; then she hastily tucked away her treasure, and stepped discreetly down into the soft harrowed and hoed land, and began to drop corn again, five kernels to a hill. She had seen the top of Peggy Bond's head over the knoll, and now Peggy herself came entirely into view, gazing upward to the skies, and stumbling more or less, but counting the corn by touch and twisting her head about anxiously to gain advantage over her uncertain vision. Betsey made a friendly, inarticulate little sound as they passed; she was thinking that somebody said once that Peggy's eyesight might be remedied if she could go to Boston to the hospital; but that was so remote and impossible an undertaking that no one had ever taken the first step. Betsey Lane's brown old face suddenly worked with excitement, but in a moment more she regained her usual firm expression, and spoke carelessly to Peggy as she turned and came alongside.

The high spring wind of the morning had quite fallen; it was a lovely May afternoon. The woods about the field to the northward were full of birds, and the young leaves scarcely hid the solemn shapes of a company of crows that patiently attended the corn-planting. Two of the men had finished their hoeing, and were busy with the construction of a scarecrow; they knelt in the furrows, chuckling, and looking over some forlorn, discarded garments. It was a time-honored custom to make the scarecrow resemble one of the poor-house family; and this year they intended to have Mrs Lavina Dow protect the field in effigy; last year it was the counterfeit of Betsey Lane who stood on guard, with an easily recognized quilted hood and the remains of a valued shawl that one of the calves had found airing on a fence and chewed to pieces. Behind the men was the foundation for this rustic attempt at statuary,—an upright stake and bar in the form of a cross. This stood on the highest part of the field; and as the men knelt near it, and the quaint figures of the corn-planters went and came, the scene gave a curious suggestion of foreign life. It was not like New England; the presence of the rude cross appealed strangely to the imagination.

IV

LIFE flowed so smoothly, for the most part, at the Byfleet Poor-farm, that nobody knew what to make, later in the summer, of a strange disappearance. All the elder inmates were familiar with illness and death, and the poor pomp of a town-pauper's funeral. The comings and goings and the various misfortunes of those who composed this strange family, related only through its disasters, hardly served for the excitement and talk of a single day. Now that the June days were at their longest, the old people were sure to wake earlier than ever; but one morning, to the astonishment of every one, Betsey Lane's bed was empty; the sheets and blankets, which were her own, and guarded with jealous care, were carefully folded and placed on a chair not too near the window, and Betsey had flown. Nobody had heard her go down the creaking stairs. The kitchen door was unlocked, and the old watchdog lay on the step outside in the early sunshine, wagging his tail and looking wise, as if he were left on guard and meant to keep the fugitive's secret.

'Never knowed her to do nothin' afore 'thout talking it over a fortnight, and paradin' off when we could all see her,' ventured a spiteful voice. 'Guess we can wait till night to hear 'bout it.'

Mrs Dow looked sorrowful and shook her head. 'Betsey had an aunt on her mother's side that went and drownded of herself; she was a pritty-appearing woman as ever you see.'

'Perhaps she's gone to spend the day with Decker's folks,' suggested Peggy Bond. 'She always takes an extra early start; she was speakin' lately o' going up their way;' but Mrs Dow shook her head with a most melancholy look. 'I'm impressed that something's befell her,' she insisted. 'I heard her a-groanin' in her sleep. I was wakeful the forepart o' the night,—'t is very unusual with me, too.'

' 'T wa'n't like Betsey not to leave us any word,' said the other old friend, with more resentment than melancholy. They sat together almost in silence that morning in the shed chamber. Mrs Dow was sorting and cutting rags, and Peggy braided them into long ropes, to be made into mats at a later date. If they had only known where Betsey Lane had gone, they might have talked about it until dinner-time at noon; but failing this new subject, they could take no interest in any of their old ones. Out in the field the corn was well up, and the men were hoeing. It was a hot morning in the shed chamber, and the woolen rags were dusty and hot to handle.

V

BYFLEET people knew each other well, and when this mysteriously absent person did not return to the town-farm at the end of a week, public interest became much excited; and presently it was ascertained that Betsey Lane was neither making a visit to her friends the Deckers on Birch Hill, nor to any nearer acquaintances; in fact, she had disappeared altogether from her wonted haunts. Nobody remembered to have seen her pass, hers had been such an early flitting; and when somebody thought of her having gone away by train, he was laughed at for forgetting that the earliest morning train from South Byfleet, the nearest station, did not start until long after eight o'clock; and if Betsey had designed to be one of the passengers, she would have started along the road at seven, and been seen and known of all women. There was not a kitchen in that part of Byfleet that did not have windows toward the road. Conversation rarely left the level of the neighborhood gossip: to see Betsey Lane, in her best clothes, at that hour in the morning, would have been the signal for much exercise of imagination; but as day after day went by without news, the curiosity of those who knew her best turned slowly into fear, and at last Peggy Bond again gave utterance to the belief that Betsey had either gone out in the early morning and put an end to her life, or that she had gone to the Centennial. Some of the people at table were moved to loud laughter,—it was at supper-time on a Sunday night,— but others listened with great interest.

'She never'd put on her good clothes to drownd herself,' said the widow. 'She might have thought 't was good as takin' 'em with her, though. Old folks has wandered off an' got lost in the woods afore now.'

Mrs Dow and Peggy resented this impertinent remark, but deigned to take no notice of the speaker. 'She wouldn't have wore her best clothes to the Centennial, would she?' mildly inquired Peggy, bobbing her head toward the ceiling. ' 'T would be a shame to spoil your best things in such a place. An' I don't know of her havin' any money; there's the end o' that.'

'You're bad as old Mis' Bland, that used to live neighbor to our folks,' said one of the old men. 'She was dreadful precise; an' she so begretched* to wear a good alapaca dress that was left to her, that it hung in a press forty year, an' baited the moths at last.'

'I often seen Mis' Bland a-goin' in to meetin' when I was a young girl,' said Peggy Bond approvingly. 'She was a good-appearin' woman, an' she left property.'

'Wish she'd left it to me, then,' said the poor soul opposite, glancing at her pathetic row of children: but it was not good manners at the farm to deplore one's situation, and Mrs Dow and Peggy only frowned. 'Where do you suppose Betsey can be?' said Mrs Dow, for the twentieth time. 'She didn't have no money. I know she ain't gone far, if it's so that she's yet alive. She's b'en real pinched all the spring.'

'Perhaps that lady that come one day give her some,' the keeper's wife suggested mildly.

'Then Betsey would have told me,' said Mrs Dow, with injured dignity.

VI

ON the morning of her disappearance, Betsey rose even before the pewee* and the English sparrow, and dressed herself quietly, though with trembling hands, and stole out of the kitchen door like a plunderless thief. The old dog licked her hand and looked at her anxiously; the tortoise-shell cat rubbed against her best gown, and trotted away up the yard, then she turned anxiously and came after the old woman, following faithfully until she had to be driven back. Betsey was used to long country excursions afoot. She dearly loved the early morning; and finding that there was no dew to trouble her, she began to follow pasture paths and short cuts across the fields, surprising here and there a flock of sleepy sheep, or a startled calf that rustled out from the bushes. The birds were pecking their breakfast from bush and turf; and hardly any of the wild inhabitants of that rural world were enough alarmed by her presence to do more than flutter away if they chanced to be in her path. She stepped along, light-footed and eager as a girl, dressed in her neat old straw bonnet and black gown, and carrying a few belongings in her best bundle-handkerchief, one that her only brother had brought home from the East Indies fifty years before. There was an old crow perched as sentinel on a small, dead pine-tree, where he could warn friends who were pulling up the sprouted corn in a field close by; but he only gave a contemptuous caw as the adventurer appeared, and she shook her bundle at him in revenge, and laughed to see him so clumsy as he tried to keep his footing on the twigs.

'Yes, I be,' she assured him. 'I'm a-goin' to Pheladelphy, to the Centennial, same 's other folks. I'd jest as soon tell ye's not, old crow;' and Betsey laughed aloud in pleased content with herself and her daring, as she walked along. She had only two miles to go to the station at South Byfleet, and she felt for the money now and then, and found it safe enough. She took great pride in the success of her escape, and especially in the long concealment of her wealth. Not a night had passed since Mrs Strafford's visit that she had not slept with the roll of money under her pillow by night, and buttoned safe inside her dress by day. She knew that everybody would offer advice and even commands about the spending or saving of it; and she brooked no interference.

The last mile of the foot-path to South Byfleet was along the railway track; and Betsey began to feel in haste, though it was still nearly two hours to train time. She looked anxiously forward and back along the rails every few minutes, for fear of being run over; and at last she caught sight of an engine that was apparently coming toward her, and took flight into the woods before she could gather courage to follow the path again. The freight train proved to be at a standstill, waiting at a turnout; and some of the men were straying about, eating their early breakfast comfortably in this time of leisure. As the old woman came up to them, she stopped too, for a moment of rest and conversation.

'Where be ye goin'?' she asked pleasantly; and they told her. It was to the town where she had to change cars and take the great through train; a point of geography which she had learned from evening talks between the men at the farm.

'What'll ye carry me there for?'

'We don't run no passenger cars,' said one of the young fellows, laughing. 'What makes you in such a hurry?'

'I'm startin' for Pheladelphy, an' it's a gre't ways to go.'

'So 't is; but you're consid'able early, if you're makin' for the eight-forty train. See here! you haven't got a needle an' thread 'long of you in that bundle, have you? If you'll sew me on a couple o' buttons, I'll give ye a free ride. I'm in a sight o' distress, an' none o' the fellows is provided with as much as a bent pin.'

'You poor boy! I'll have you seen to, in half a minute. I'm troubled with a stiff arm, but I'll do the best I can.'

The obliging Betsey seated herself stiffly on the slope of the embankment, and found her thread and needle with utmost haste. Two of the train-men stood by and watched the careful stitches, and even offered her a place as spare brakeman, so that they might keep her near; and Betsey took the offer with considerable seriousness, only thinking it necessary to assure them that she was getting most too old to be out in all weathers. An express went by like an earthquake, and she was presently hoisted on board an empty box-car by two of her new and flattering acquaintances, and found herself before noon at the end of the first stage of her journey, without having spent a cent, and furnished with any amount of thrifty advice. One of the young men, being compassionate of her unprotected state as a traveler, advised her to find out the widow of an uncle of his in Philadelphia, saying despairingly that he couldn't tell her just how to find the house; but Miss Betsey Lane said that she had an English tongue in her head, and should be sure to find whatever she was looking for. This unexpected incident of the freight train was the reason why everybody about the South Byfleet station insisted that no such person had taken passage by the regular train that same morning, and why there were those who persuaded themselves that Miss Betsey Lane was probably lying at the bottom of the poor-farm pond.

VII

'LAND sakes!' said Miss Betsey Lane, as she watched a Turkish person parading by in his red fez, 'I call the Centennial somethin' like the day o' judgment!* I wish I was goin' to stop a month, but I dare say 't would be the death o' my poor old bones.'

She was leaning against the barrier of a patent pop-corn establishment, which had given her a sudden reminder of home, and of the winter nights when the sharp-kerneled little red and yellow ears were brought out, and Old Uncle Eph Flanders sat by the kitchen stove, and solemnly filled a great wooden chopping-tray for the refreshment of the company. She had wandered and loitered and looked until her eyes and head had grown numb and unreceptive; but it is only unimaginative persons who can be really astonished. The imagination can always outrun the possible and actual sights and sounds of the world; and this plain old body from Byfleet rarely found anything rich and splendid enough to surprise her. She saw the wonders of the West and the splendors of the East with equal calmness and satisfaction; she had always known that there was an amazing world outside the boundaries of Byfleet. There was a piece of paper in her pocket on which was marked, in her clumsy handwriting, 'If Betsey Lane should meet with accident, notify the selectmen of Byfleet;' but having made this slight provision for the future, she had thrown herself boldly into the sea of strangers, and then had made the joyful discovery that friends were to be found at every turn.

There was something delightfully companionable about Betsey; she had a way of suddenly looking up over her big spectacles with a reassuring and expectant smile, as if you were going to speak to her, and you generally did. She must have found out where hundreds of people came from, and whom they had left at home, and what they thought of the great show, as she sat on a bench to rest, or leaned over the railings where free luncheons were afforded by the makers of hot waffles and molasses candy and fried potatoes; and there was not a night when she did not return to her lodgings with a pocket crammed with samples of spool cotton and nobody knows what. She had already collected small presents for almost everybody she knew at home, and she was such a pleasant, beaming old country body, so unmistakably appreciative and interested, that nobody ever thought of wishing that she would move on. Nearly all the busy people of the Exhibition called her either Aunty or Grandma at once, and made little pleasures for her as best they could. She was a delightful contrast to the indifferent, stupid crowd that drifted along, with eyes fixed at the same level, and seeing, even on that level, nothing for fifty feet at a time. 'What be you making here, dear?' Betsey Lane would ask joyfully, and the most perfunctory guardian hastened to explain. She squandered money as she had never had the pleasure of doing before, and this hastened the day when she must return to Byfleet. She was always inquiring if there were any spectacle-sellers at hand, and received occasional directions; but it was a difficult place for her to find her way about in, and the very last day of her stay arrived before she found an exhibitor of the desired sort, an oculist and instrument-maker.

'I called to get some specs for a friend that's upsighted,' she gravely informed the salesman, to his extreme amusement. 'She's dreadful troubled, and jerks her head up like a hen a-drinkin'. She's got a blur a-growin' an' spreadin', an' sometimes she can see out to one side on 't, and more times she can't.'

'Cataracts,' said a middle-aged gentleman at her side; and Betsey Lane turned to regard him with approval and curiosity.

' 'T is Miss Peggy Bond I was mentioning, of Byfleet Poor-farm,' she explained. 'I count on gettin' some glasses to relieve her trouble, if there's any to be found.'

'Glasses won't do her any good,' said the stranger. 'Suppose you come and sit down on this bench, and tell me all about it. First, where is Byfleet?' and Betsey gave the directions at length.

'I thought so,' said the surgeon. 'How old is this friend of yours?'

Betsey cleared her throat decisively, and smoothed her gown over her knees as if it were an apron; then she turned to take a good look at her new acquaintance as they sat on the rustic bench together. 'Who be you, sir, I should like to know?' she asked, in a friendly tone.

'My name's Dunster.'

'I take it you're a doctor,' continued Betsey, as if they had overtaken each other walking from Byfleet to South Byfleet on a summer morning.

'I'm a doctor; part of one at least,' said he. 'I know more or less about eyes; and I spend my summers down on the shore at the mouth of your river; some day I'll come up and look at this person. How old is she?'

'Peggy Bond is one that never tells her age; 't ain't come quite up to where she'll begin to brag of it, you see,' explained Betsey reluctantly; 'but I know her to be nigh to seventy-six, one way or t' other. Her an' Mrs Mary Ann Chick was same year's child'n, and Peggy knows I know it, an' two or three times when we've be'n in the buryin'-ground where Mary Ann lays an' has her dates right on her headstone, I couldn't bring Peggy to take no sort o' notice. I will say she makes, at times, a convenience of being upsighted. But there, I feel for her,—everybody does; it keeps her stubbin' an' trippin' against everything, beakin'* and gazin' up the way she has to.'

'Yes, yes,' said the doctor, whose eyes were twinkling. 'I'll come and look after her, with your town doctor, this summer,—some time in the last of July or first of August.'

'You'll find occupation,' said Betsey, not without an air of patronage. 'Most of us to the Byfleet Farm has got our ails, now I tell ye. You ain't got no bitters that'll take a dozen years right off an ol' lady's shoulders?'

The busy man smiled pleasantly, and shook his head as he went away. 'Dunster,' said Betsey to herself, soberly committing the new name to her sound memory. 'Yes, I mustn't forget to speak of him to the doctor, as he directed. I do' know now as Peggy would vally herself quite so much accordin' to, if she had her eyes fixed same as other folks. I expect there wouldn't been a smarter woman in town, though, if she'd had a proper chance. Now I've done what I set to do for her, I do believe, an' 't wa'n't glasses, neither. I'll git her a pritty little shawl with that money I laid aside. Peggy Bond ain't got a pritty shawl. I always wanted to have a real good time, an' now I'm havin' it.'

VIII

Two or three days later, two pathetic figures might have been seen crossing the slopes of the poor-farm field, toward the low shores of Byfield pond. It was early in the morning, and the stubble of the lately mown grass was wet with rain and hindering to old feet. Peggy Bond was more blundering and liable to stray in the wrong direction than usual; it was one of the days when she could hardly see at all. Aunt Lavina Dow was unusually clumsy of movement, and stiff in the joints; she had not been so far from the house for three years. The morning breeze filled the gathers of her wide gingham skirt, and aggravated the size of her unwieldy figure. She supported herself with a stick, and trusted beside to the fragile support of Peggy's arm. They were talking together in whispers.

'Oh, my sakes!' exclaimed Peggy, moving her small head from side to side. 'Hear you wheeze, Mis' Dow! This may be the death o' you; there, do go slow! You set here on the side-hill, an' le' me go try if I can see.'

'It needs more eyesight than you've got,' said Mrs Dow, panting between the words. 'Oh! to think how spry I was in my young days, an' here I be now, the full of a door, an' all my complaints so aggravated by my size. 'T is hard! 't is hard! but I'm a-doin' of all this for pore Betsey's sake. I know they've all laughed, but I look to see her ris' to the top o' the pond this day,—'t is just nine days since she departed; an' say what they may, I know she hove herself in. It run in her family; Betsey had an aunt that done just so, an' she ain't be'n like herself, a-broodin' an' hivin' away alone, an' nothin' to say to you an' me that was always sich good company all together. Somethin' sprung her mind, now I tell ye, Mis' Bond.'

'I feel to hope we sha'n't find her, I must say,' faltered Peggy. It was plain that Mrs Dow was the captain of this doleful expedition. 'I guess she ain't never thought o' drowndin' of herself, Mis' Dow; she's gone off a-visitin' way over to the other side o' South Byfleet; some thinks she's gone to the Centennial even now!'

'She hadn't no proper means, I tell ye,' wheezed Mrs Dow indignantly; 'an' if you prefer that others should find her floatin' to the top this day, instid of us that's her best friends, you can step back to the house.'

They walked on in aggrieved silence. Peggy Bond trembled with excitement, but her companion's firm grasp never wavered, and so they came to the narrow, gravelly margin and stood still. Peggy tried in vain to see the glittering water and the pond-lilies that starred it; she knew that they must be there; once, years ago, she had caught fleeting glimpses of them, and she never forgot what she had once seen. The clear blue sky overhead, the dark pine-woods beyond the pond, were all clearly pictured in her mind. 'Can't you see nothin'?' she faltered; 'I believe I'm wuss'n upsighted this day. I'm going to be blind.'

'No,' said Lavina Dow solemnly; 'no, there ain't nothin' whatever, Peggy. I hope to mercy she ain't'—

'Why, whoever'd expected to find you 'way out here!' exclaimed a brisk and cheerful voice. There stood Betsey Lane herself, close behind them, having just emerged from a thicket of alders that grew close by. She was following the short way homeward from the railroad.

'Why, what's the matter, Mis' Dow? You ain't overdoin', be ye? an' Peggy's all of a flutter. What in the name o' natur' ails ye?'

'There ain't nothin' the matter, as I knows on,' responded the leader of this fruitless expedition. 'We only thought we'd take a stroll this pleasant mornin',' she added, with sublime self-possession. 'Where've you be'n, Betsey Lane?'

'To Pheladelphy, ma'am,' said Betsey, looking quite young and gay, and wearing a townish and unfamiliar air that upheld her words. 'All ought to go that can; why, you feel 's if you'd be'n all round the world. I guess I've got enough to think of and tell ye for the rest o' my days. I've always wanted to go somewheres. I wish you'd be'n there, I do so. I've talked with folks from Chiny an' the back o' Pennsylvany; and I see folks way from Australy that 'peared as well as anybody; an' I see how they made spool cotton, an' sights o' other things; an' I spoke with a doctor that lives down to the beach in the summer, an' he offered to come up 'long in the first of August, an' see what he can do for Peggy's eyesight. There was di'monds there as big as pigeon's eggs; an' I met with Mis' Abby Fletcher from South Byfleet depot; an' there was hogs there that weighed risin' thirteen hunderd'—

'I want to know,' said Mrs Lavina Dow and Peggy Bond, together.

'Well, 't was a great exper'ence for a person,' added Lavina, turning ponderously, in spite of herself, to give a last wistful look at the smiling waters of the pond.

'I don't know how soon I be goin' to settle down,' proclaimed the rustic sister of Sindbad.* 'What's for the good o' one's for the good of all. You just wait till we're setting together up in the old shed chamber! You know, my dear Mis' Katy Strafford give me a han'some present o' money that day she come to see me; and I'd be'n a-dreamin' by night an' day o' seein' that Centennial; and when I come to think on 't I felt sure somebody ought to go from this neighborhood, if 't was only for the good o' the rest; and I thought I'd better be the one. I wa'n't goin' to ask the selec'men neither. I've come back with one-thirty-five in money, and I see everything there, an' I fetched ye all a little somethin'; but I'm full o' dust now, an' pretty nigh beat out. I never see a place more friendly than Pheladelphy; but 't ain't natural to a Byfleet person to be always walkin' on a level. There, now, Peggy, you take my bundle-handkercher and the basket, and let Mis' Dow sag on to me. I'll git her along twice as easy.'

With this the small elderly company set forth triumphant toward the poor-house, across the wide green field.

THE ONLY ROSE

I

JUST where the village abruptly ended, and the green mowing fields began, stood Mrs Bickford's house, looking down the road with all its windows, and topped by two prim chimneys that stood up like ears. It was placed with an end to the road, and fronted southward; you could follow a straight path from the gate past the front door and find Mrs Bickford sitting by the last window of all in the kitchen, unless she were solemnly stepping about, prolonging the stern duties of her solitary housekeeping.

One day in early summer, when almost every one else in Fairfield had put her house plants out of doors, there were still three flower pots on a kitchen window sill. Mrs Bickford spent but little time over her rose and geranium and Jerusalem cherry-tree, although they had gained a kind of personality born of long association. They rarely undertook to bloom but had most courageously maintained life in spite of their owner's unsympathetic but conscientious care. Later in the season she would carry them out of doors, and leave them, until the time of frosts, under the shade of a great apple-tree, where they might make the best of what the summer had to give.

The afternoon sun was pouring in, the Jerusalem cherry-tree drooped its leaves in the heat and looked pale, when a neighbor, Miss Pendexter, came in from the next house but one to make a friendly call. As she passed the parlor with its shut blinds, and the sitting-room, also shaded carefully from the light, she wished, as she had done many times before, that somebody beside the owner might have the pleasure of living in and using so good and pleasant a house. Mrs Bickford always complained of having so much care, even while she valued herself intelligently upon having the right to do as she pleased with one of the best houses in Fairfield. Miss Pendexter was a cheerful, even gay little person, who always brought a pleasant flurry of excitement, and usually had a genuine though small piece of news to tell, or some new aspect of already received information.

Mrs Bickford smiled as she looked up to see this sprightly neighbor coming. She had no gift at entertaining herself, and was always glad, as one might say, to be taken off her own hands.

Miss Pendexter smiled back, as if she felt herself to be equal to the occasion.

'How be you to-day?' the guest asked kindly, as she entered the kitchen. 'Why, what a sight o' flowers, Mis' Bickford! What be you goin' to do with 'em all?'

Mrs Bickford wore a grave expression as she glanced over her spectacles. 'My sister's boy fetched 'em over,' she answered. 'You know my sister Parsons's a great hand to raise flowers, an' this boy takes after her. He said his mother thought the gardin never looked handsomer, and she picked me these to send over. They was sendin' a team to Westbury for some fertilizer to put on the land, an' he come with the men, an' stopped to eat his dinner 'long o' me. He's been growin' fast, and looks peakèd. I expect sister 'Liza thought the ride, this pleasant day, would do him good. 'Liza sent word for me to come over and pass some days next week, but it ain't so that I can.'

'Why, it's a pretty time of year to go off and make a little visit,' suggested the neighbor encouragingly.

'I ain't got my sitting-room chamber carpet taken up yet,' sighed Mrs Bickford. 'I do feel condemned. I might have done it to-day, but 't was all at end when I saw Tommy coming. There, he's a likely boy, an' so relished his dinner; I happened to be well prepared. I don't know but he's my favorite o' that family. Only I've been sittin' here thinkin', since he went, an' I can't remember that I ever was so belated with my spring cleaning.'

''T was owin' to the weather,' explained Miss Pendexter. 'None of us could be so smart as common this year, not even the lazy ones that always get one room done the first o' March, and brag of it to others' shame, and then never let on when they do the rest.'

The two women laughed together cheerfully. Mrs Bickford

had put up the wide leaf of her large table between the windows and spread out the flowers. She was sorting them slowly into three heaps.

'Why, I do declare if you haven't got a rose in bloom yourself!' exclaimed Miss Pendexter abruptly, as if the bud had not been announced weeks before, and its progress regularly commented upon. 'Ain't it a lovely rose? Why, Mis' Bickford!'

'Yes 'm, it's out to-day,' said Mrs Bickford, with a somewhat plaintive air. 'I'm glad you come in so as to see it.'

The bright flower was like a face. Somehow, the beauty and life of it were surprising in the plain room, like a gay little child who might suddenly appear in a doorway. Miss Pendexter forgot herself and her hostess and the tangled mass of garden flowers in looking at the red rose. She even forgot that it was incumbent upon her to carry forward the conversation. Mrs Bickford was subject to fits of untimely silence which made her friends anxiously sweep the corners of their minds in search of something to say, but any one who looked at her now could easily see that it was not poverty of thought that made her speechless, but an overburdening sense of the inexpressible.

'Goin' to make up all your flowers into bo'quets? I think the short-stemmed kinds is often pretty in a dish,' suggested Miss Pendexter compassionately.

'I thought I should make them into three bo'quets. I wish there wa'n't quite so many. Sister Eliza's very lavish with her flowers; she's always been a kind sister, too,' said Mrs Bickford vaguely. She was not apt to speak with so much sentiment, and as her neighbor looked at her narrowly she detected unusual signs of emotion. It suddenly became evident that the three nosegays were connected in her mind with her bereavement of three husbands, and Miss Pendexter's easily roused curiosity was quieted by the discovery that her friend was bent upon a visit to the burying-ground. It was the time of year when she was pretty sure to spend an afternoon there, and sometimes they had taken the walk in company. Miss Pendexter expected to receive the usual invitation, but there was nothing further said at the moment, and she looked again at the pretty rose.

Mrs Bickford aimlessly handled the syringas and flowering almond sprays, choosing them out of the fragrant heap only to lay them down again. She glanced out of the window; then gave Miss Pendexter a long expressive look.

'I expect you're going to carry 'em over to the burying-ground?' inquired the guest, in a sympathetic tone.

'Yes 'm,' said the hostess, now well started in conversation and in quite her every-day manner. 'You see I was goin' over to my brother's folks to-morrow in South Fairfield, to pass the day; they said they were goin' to send over to-morrow to leave a wagon at the blacksmith's, and they'd hitch that to their best chaise, so I could ride back very comfortable. You know I have to avoid bein' out in the mornin' sun?'

Miss Pendexter smiled to herself at this moment; she was obliged to move from her chair at the window, the May sun was so hot on her back, for Mrs Bickford always kept the curtains rolled high up, out of the way, for fear of fading and dust. The kitchen was a blaze of light. As for the Sunday chaise being sent, it was well known that Mrs Bickford's married brothers and sisters comprehended the truth that she was a woman of property, and had neither chick nor child.

'So I thought 't was a good opportunity to just stop an' see if the lot was in good order,—last spring Mr Wallis's stone hove with the frost; an' so I could take these flowers.' She gave a sigh. 'I ain't one that can bear flowers in a close room,— they bring on a headache; but I enjoy 'em as much as anybody to look at, only you never know what to put 'em in. If I could be out in the mornin' sun, as some do, and keep flowers in the house, I should have me a gardin, certain,' and she sighed again.

'A garden's a sight o' care, but I don't begrudge none o' the care I give to mine. I have to scant on flowers so 's to make room for pole beans,' said Miss Pendexter gayly. She had only a tiny strip of land behind her house, but she always had something to give away, and made riches out of her narrow poverty. 'A few flowers gives me just as much pleasure as more would,' she added. 'You get acquainted with things when you've only got one or two roots. My sweet-williams is just like folks.'

'Mr Bickford was partial to sweet-williams,' said Mrs Bickford. 'I never knew him to take notice of no other sort of
flowers. When we'd be over to Eliza's, he'd walk down her
gardin, an' he'd never make no comments until he come to
them, and then he'd say, "Those is sweet-williams." How many
times I've heard him!'

'You ought to have a sprig of 'em for his bo'quet,'
suggested Miss Pendexter.

'Yes, I've put a sprig in,' said her companion.

At this moment Miss Pendexter took a good look at the
bouquets, and found that they were as nearly alike as careful
hands could make them. Mrs Bickford was evidently trying to
reach absolute impartiality.

'I don't know but you think it's foolish to tie 'em up this
afternoon,' she said presently, as she wound the first with a
stout string. 'I thought I could put 'em in a bucket o' water
out in the shed, where there's a draught o' air, and then I
should have all my time in the morning. I shall have a good
deal to do before I go. I always sweep the setting-room and
front entry Wednesdays. I want to leave everything nice, goin'
away for all day so. So I meant to get the flowers out o' the way
this afternoon. Why, it's most half past four, ain't it? But I
sha'n't pick the rose till mornin'; 't will be blowed out better
then.'

'The rose?' questioned Miss Pendexter. 'Why, are you goin'
to pick that, too?'

'Yes, I be. I never like to let 'em fade on the bush. There,
that's just what's a-troublin' me,' and she turned to give a
long, imploring look at the friend who sat beside her. Miss
Pendexter had moved her chair before the table in order to
be out of the way of the sun. 'I don't seem to know which of
'em ought to have it,' said Mrs Bickford despondently. 'I do so
hate to make a choice between 'em; they all had their good
points, especially Mr Bickford, and I respected 'em all. I don't
know but what I think of one on 'em 'most as much as I do of
the other.'

'Why, 't is difficult for you, ain't it?' responded Miss Pendexter. 'I don't know 's I can offer advice.'

'No, I s'pose not,' answered her friend slowly, with a

shadow of disappointment coming over her calm face. 'I feel sure you would if you could, Abby.'

Both of the women felt as if they were powerless before a great emergency.

'There's one thing,—they're all in a better world now,' said Miss Pendexter, in a self-conscious and constrained voice; 'they can't feel such little things or take note o' slights same 's we can.'

'No; I suppose 't is myself that wants to be just,' answered Mrs Bickford. 'I feel under obligations to my last husband when I look about and see how comfortable he left me. Poor Mr Wallis had his great projects, an' perhaps if he'd lived longer he'd have made a record; but when he died he'd failed all up, owing to that patent corn-sheller he'd put everything into, and, as you know, I had to get along 'most any way I could for the next few years. Life was very disappointing with Mr Wallis, but he meant well, an' used to be an amiable person to dwell with, until his temper got spoilt makin' so many hopes an' havin' 'em turn out failures. He had consider'ble of an air, an' dressed very handsome when I was first acquainted with him, Mr Wallis did. I don't know 's you ever knew Mr Wallis in his prime?'

'He died the year I moved over here from North Denfield,' said Miss Pendexter, in a tone of sympathy. 'I just knew him by sight. I was to his funeral. You know you lived in what we call the Wells house then, and I felt it wouldn't be an intrusion, we was such near neighbors. The first time I ever was in your house was just before that, when he was sick, an' Mary 'Becca Wade an' I called to see if there was anything we could do.'

'They used to say about town that Mr Wallis went to an' fro like a mail-coach an' brought nothin' to pass,' announced Mrs Bickford without bitterness. 'He ought to have had a better chance than he did in this little neighborhood. You see, he had excellent ideas, but he never 'd learned the machinist's trade, and there was somethin' the matter with every model he contrived. I used to be real narrow-minded when he talked about moving 'way up to Lowell, or some o' them places; I hated to think of leaving my folks; and now I see that I never done right by him. His ideas was good. I know once he was on a jury, and there was a man stopping to the tavern where he was, near the court house, a man that traveled for a firm to Lowell; and they engaged in talk, an' Mr Wallis let out some o' his notions an' contrivances, an' he said that man wouldn't hardly stop to eat, he was so interested, an' said he'd look for a chance for him up to Lowell. It all sounded so well that I kind of begun to think about goin' myself. Mr Wallis said we'd close the house here, and go an' board through the winter. But he never heard a word from him, and the disappointment was one he never got over. I think of it now different from what I did then. I often used to be kind of disapproving to Mr Wallis; but there, he used to be always tellin' over his great projects. Somebody told me once that a man by the same name of the one he met while he was to court had got some patents for the very things Mr Wallis used to be workin' over; but 't was after he died, an' I don't know 's 't was in him to ever really set things up so other folks could ha' seen their value. His machines always used to work kind of rickety, but folks used to come from all round to see 'em; they was curiosities if they wa'n't nothin' else, an' gave him a name.'

Mrs Bickford paused a moment, with some geranium leaves in her hand, and seemed to suppress with difficulty a desire to speak even more freely.

'He was a dreadful notional man,' she said at last, regretfully, and as if this fact were a poor substitute for what had just been in her mind. 'I recollect one time he worked all through the early winter over my churn, an' got it so it would go three quarters of an hour all of itself if you wound it up; an' if you'll believe it, he went an' spent all that time for nothin' when the cow was dry, an' we was with difficulty borrowin' a pint o' milk a day somewhere in the neighborhood just to get along with.' Mrs Bickford flushed with displeasure, and turned to look at her visitor. 'Now what do you think of such a man as that, Miss Pendexter?' she asked.

'Why, I don't know but 't was just as good for an invention,' answered Miss Pendexter timidly; but her friend looked doubtful, and did not appear to understand.

'Then I asked him where it was, one day that spring when

I'd got tired to death churnin', an' the butter wouldn't come
in a churn I'd had to borrow, and he'd gone an' took ours all
to pieces to get the works to make some other useless contriv-
ance with. He had no sort of a business turn, but he was well
meanin', Mr Wallis was, an' full o' divertin' talk; they used to
call him very good company. I see now that he never had no
proper chance. I've always regretted Mr Wallis,' said she who
was now the widow Bickford.

'I'm sure you always speak well of him,' said Miss Pendexter. ' 'T was a pity he hadn't got among good business men,
who could push his inventions an' do all the business part.'

'I was left very poor an' needy for them next few years,' said
Mrs Bickford mournfully; 'but he never 'd give up but what he
should die worth his fifty thousand dollars. I don't see now
how I ever did get along them next few years without him; but
there, I always managed to keep a pig, an' sister Eliza gave me
my potatoes, and I made out somehow. I could dig me a few
greens, you know, in spring, and then 't would come
strawberry-time, and other berries a-followin' on. I was always
decent to go to meetin' till within the last six months, an' then
I went in bad weather, when folks wouldn't notice; but 't was
a rainy summer, an' I managed to get considerable preachin'
after all. My clothes looked proper enough when 't was a wet
Sabbath. I often think o' them pinched days now, when I'm
left so comfortable by Mr Bickford.'

'Yes 'm, you've everything to be thankful for,' said Miss
Pendexter, who was as poor herself at that moment as her
friend had ever been, and who could never dream of ventur-
ing upon the support and companionship of a pig. 'Mr Bickford was a very personable man,' she hastened to say, the
confidences were so intimate and interesting.

'Oh, very,' replied Mrs Bickford; 'there was something
about him that was very marked. Strangers would always ask
who he was as he come into meetin'. His words counted; he
never spoke except he had to. 'T was a relief at first after Mr
Wallis's being so fluent; but Mr Wallis was splendid company
for winter evenings,—'t would be eight o'clock before you
knew it. I didn't use to listen to it all, but he had a great deal
of information. Mr Bickford was dreadful dignified; I used to

be sort of meechin'* with him along at the first, for fear he'd disapprove of me; but I found out 't wa'n't no need; he was always just that way, an' done everything by rule an' measure. He hadn't the mind of my other husbands, but he was a very dignified appearing man; he used 'most always to sleep in the evenin's, Mr Bickford did.'

'Them is lovely bo'quets, certain!' exclaimed Miss Pendexter. 'Why, I couldn't tell 'em apart; the flowers are comin' out just right, aren't they?'

Mrs Bickford nodded assent, and then, startled by sudden recollection, she cast a quick glance at the rose in the window.

'I always seem to forget about your first husband, Mr Fraley,' Miss Pendexter suggested bravely. 'I've often heard you speak of him, too, but he'd passed away long before I ever knew you.'

'He was but a boy,' said Mrs Bickford. 'I thought the world was done for me when he died, but I've often thought since 't was a mercy for him. He come of a very melancholy family, and all his brothers an' sisters enjoyed poor health; it might have been his lot. Folks said we was as pretty a couple as ever come into church; we was both dark, with black eyes an' a good deal o' color,—you wouldn't expect it to see me now. Albert was one that held up his head, and looked as if he meant to own the town, an' he had a good word for everybody. I don't know what the years might have brought.'

There was a long pause. Mrs Bickford leaned over to pick up a heavy-headed Guelder-rose that had dropped on the floor.

'I expect 't was what they call fallin' in love,' she added, in a different tone; 'he wa'n't nothin' but a boy, an' I wa'n't nothin' but a girl, but we was dreadful happy. He didn't favor his folks,—they all had hay-colored hair and was faded-looking, except his mother; they was alike, and looked alike, an' set everything by each other. He was just the kind of strong, hearty young man that goes right off if they get a fever. We was just settled on a little farm, an' he'd have done well if he'd had time; as it was, he left debts. He had a hasty temper, that was his great fault, but Albert had a lovely voice to sing; they said there wa'n't no such tenor voice in this part o' the State.

I could hear him singin' to himself right out in the field a-ploughin' or hoein', an' he didn't know it half o' the time, no more 'n a common bird would. I don't know 's I valued his gift as I ought to, but there was nothin' ever sounded so sweet to me. I ain't one that ever had much fancy, but I knowed Albert had a pretty voice.'

Mrs Bickford's own voice trembled a little, but she held up the last bouquet and examined it critically. 'I must hurry now an' put these in water,' she said, in a matter of fact tone. Little Miss Pendexter was so quiet and sympathetic that her hostess felt no more embarrassed than if she had been talking only to herself.

'Yes, they do seem to droop some; 't is a little warm for them here in the sun,' said Miss Pendexter; 'but you'll find they'll all come up if you give them their fill o' water. They'll look very handsome to-morrow; folks 'll notice them from the road. You've arranged them very tasty, Mis' Bickford.'

'They do look pretty, don't they?' Mrs Bickford regarded the three in turn. 'I want to have them all pretty. You may deem it strange, Abby.'

'Why, no, Mis' Bickford,' said the guest sincerely, although a little perplexed by the solemnity of the occasion. 'I know how 't is with friends,—that having one don't keep you from wantin' another; 't is just like havin' somethin' to eat, and then wantin' somethin' to drink just the same. I expect all friends find their places.'

But Mrs Bickford was not interested in this figure, and still looked vague and anxious as she began to brush the broken stems and wilted leaves into her wide calico apron. 'I done the best I could while they was alive,' she said, 'and mourned 'em when I lost 'em, an' I feel grateful to be left so comfortable now when all is over. It seems foolish, but I'm still at a loss about that rose.'

'Perhaps you'll feel sure when you first wake up in the morning,' answered Miss Pendexter solicitously. 'It's a case where I don't deem myself qualified to offer you any advice. But I'll say one thing, seeing 's you've been so friendly spoken and confiding with me. I never was married myself, Mis' Bickford, because it wa'n't so that I could have the one I liked.'

'I suppose he ain't livin', then? Why, I wan't never aware you had met with a disappointment, Abby,' said Mrs Bickford instantly. None of her neighbors had ever suspected little Miss Pendexter of a romance.

'Yes 'm, he's livin','' replied Miss Pendexter humbly. 'No 'm, I never have heard that he died.'

'I want to know!' exclaimed the woman of experience. 'Well, I'll tell you this, Abby: you may have regretted your lot, and felt lonesome and hardshipped, but they all have their faults, and a single woman's got her liberty, if she ain't got other blessin's.'

' 'T wouldn't have been my choice to live alone,' said Abby, meeker than before. 'I feel very thankful for my blessin's, all the same. You've always been a kind neighbor, Mis' Bickford.'

'Why can't you stop to tea?' asked the elder woman, with unusual cordiality; but Miss Pendexter remembered that her hostess often expressed a dislike for unexpected company, and promptly took her departure after she had risen to go, glancing up at the bright flower as she passed outside the window. It seemed to belong most to Albert, but she had not liked to say so. The sun was low; the green fields stretched away southward into the misty distance.

II

MRS BICKFORD's house appeared to watch her out of sight down the road, the next morning. She had lost all spirit for her holiday. Perhaps it was the unusual excitement of the afternoon's reminiscences, or it might have been simply the bright moonlight night which had kept her broad awake until dawn, thinking of the past, and more and more concerned about the rose. By this time it had ceased to be merely a flower, and had become a definite symbol and assertion of personal choice. She found it very difficult to decide. So much of her present comfort and well-being was due to Mr Bickford; still, it was Mr Wallis who had been most unfortunate, and to whom she had done least justice. If she owed recognition to Mr Bickford, she certainly owed amends to Mr Wallis. If she gave him the rose, it would be for the sake of affectionate apology. And then there was Albert, to whom she had no thought of being either indebted or forgiving. But she could not escape from the terrible feeling of indecision.

It was a beautiful morning for a drive, but Mrs Bickford was kept waiting some time for the chaise. Her nephew, who was to be her escort, had found much social advantage at the blacksmith's shop, so that it was after ten when she finally started with the three large flat-backed bouquets, covered with a newspaper to protect them from the sun. The petals of the almond flowers were beginning to scatter, and now and then little streams of water leaked out of the newspaper and trickled down the steep slope of her best dress to the bottom of the chaise. Even yet she had not made up her mind; she had stopped trying to deal with such an evasive thing as decision, and leaned back and rested as best she could.

'What an old fool I be!' she rebuked herself from time to time, in so loud a whisper that her companion ventured a respectful 'What, ma'am?' and was astonished that she made no reply. John was a handsome young man, but Mrs Bickford could never cease thinking of him as a boy. He had always been her favorite among the younger members of the family, and now returned this affectionate feeling, being possessed of an instinctive confidence in the sincerities of his prosaic aunt.

As they drove along, there had seemed at first to be something unsympathetic and garish about the beauty of the summer day. After the shade and shelter of the house, Mrs Bickford suffered even more from a contracted and assailed feeling out of doors. The very trees by the roadside had a curiously fateful, trying way of standing back to watch her, as she passed in the acute agony of indecision, and she was annoyed and startled by a bird that flew too near the chaise in a moment of surprise. She was conscious of a strange reluctance to the movement of the Sunday chaise, as if she were being conveyed against her will; but the companionship of her nephew John grew every moment to be more and more a reliance. It was very comfortable to sit by his side, even though he had nothing to say; he was manly and cheerful, and she began to feel protected.

'Aunt Bickford,' he suddenly announced, 'I may 's well out

with it! I've got a piece o' news to tell you, if you won't let on
to nobody. I expect you'll laugh, but you know I've set everything by Mary Lizzie Gifford ever since I was a boy. Well, sir!'

'Well, sir!' exclaimed aunt Bickford in her turn, quickly
roused into most comfortable self-forgetfulness. 'I am really
pleased. She'll make you a good, smart wife, John. Ain't all
the folks pleased, both sides?'

'Yes, they be,' answered John soberly, with a happy, important look that became him well.

'I guess I can make out to do something for you to help
along, when the right time comes,' said aunt Bickford impulsively, after a moment's reflection. 'I've known what it is to be
starting out in life with plenty o' hope. You ain't calculatin' on
gettin' married before fall,—or be ye?'

' 'Long in the fall,' said John regretfully. 'I wish t' we could
set up for ourselves right away this summer. I ain't got much
ahead, but I can work well as anybody, an' now I'm out o' my
time.'*

'She's a nice, modest, pretty girl. I thought she liked you,
John,' said the old aunt. 'I saw her over to your mother's, last
day I was there. Well, I expect you'll be happy.'

'Certain,' said John, turning to look at her affectionately,
surprised by this outspokenness and lack of embarrassment
between them. 'Thank you, aunt,' he said simply; 'you're a
real good friend to me;' and he looked away again hastily, and
blushed a fine scarlet over his sun-browned face. 'She's coming over to spend the day with the girls,' he added. 'Mother
thought of it. You don't get over to see us very often.'

Mrs Bickford smiled approvingly. John's mother looked for
her good opinion, no doubt, but it was very proper for John
to have told his prospects himself, and in such a pretty way.
There was no shilly-shallying about the boy.

'My gracious!' said John suddenly. 'I'd like to have drove
right by the burying-ground. I forgot we wanted to stop.'

Strange as it may appear, Mrs Bickford herself had not
noticed the burying-ground, either, in her excitement and
pleasure; now she felt distressed and responsible again, and
showed it in her face at once. The young man leaped lightly
to the ground, and reached for the flowers.

'Here, you just let me run up with 'em,' he said kindly. ' 'T is hot in the sun to-day, an' you'll mind it risin' the hill. We'll stop as I fetch you back to-night, and you can go up comfortable an' walk the yard after sundown when it's cool, an' stay as long as you're a mind to. You seem sort of tired, aunt.'

'I don't know but what I will let you carry 'em,' said Mrs Bickford slowly.

To leave the matter of the rose in the hands of fate seemed weakness and cowardice, but there was not a moment for consideration. John was a smiling fate, and his proposition was a great relief. She watched him go away with a terrible inward shaking, and sinking of pride. She had held the flowers with so firm a grasp that her hands felt weak and numb, and as she leaned back and shut her eyes she was afraid to open them again at first for fear of knowing the bouquets apart even at that distance, and giving instructions which she might regret. With a sudden impulse she called John once or twice eagerly; but her voice had a thin and piping sound, and the meditative early crickets that chirped in the fresh summer grass probably sounded louder in John's ears. The bright light on the white stones dazzled Mrs Bickford's eyes; and then all at once she felt light-hearted, and the sky seemed to lift itself higher and wider from the earth, and she gave a sigh of relief as her messenger came back along the path. 'I know who I do hope's got the right one,' she said to herself. 'There, what a touse I be in! I don't see what I had to go and pick the old rose for, anyway.'

'I declare, they did look real handsome, aunt,' said John's hearty voice as he approached the chaise. 'I set 'em up just as you told me. This one fell out, an' I kept it. I don't know 's you'll care. I can give it to Lizzie.'

He faced her now with a bright, boyish look. There was something gay in his buttonhole,—it was the red rose.

Aunt Bickford blushed like a girl. 'Your choice is easy made,' she faltered mysteriously, and then burst out laughing, there in front of the burying-ground. 'Come, get right in, dear,' she said. 'Well, well! I guess the rose was made for you; it looks very pretty in your coat, John.'

She thought of Albert, and the next moment the tears came into her old eyes. John was a lover, too.

'My first husband was just such a tall, straight young man as you be,' she said as they drove along. 'The flower he first give me was a rose.'

The Only Rose 247

She thought of Albert, and the next moment the tears came...
'My first husband was just such a tall, easy-goin' kind of a man,' she said as she drove along. 'The flower he first gave me was a rose...'

THE GUESTS OF MRS TIMMS

I

MRS PERSIS FLAGG stood in her front doorway taking leave of Miss Cynthia Pickett, who had been making a long call. They were not intimate friends. Miss Pickett always came formally to the front door and rang when she paid her visits, but, the week before, they had met at the county conference,* and happened to be sent to the same house for entertainment, and so had deepened and renewed the pleasures of acquaintance.

It was an afternoon in early June; the syringa-bushes were tall and green on each side of the stone doorsteps, and were covered with their lovely white and golden flowers. Miss Pickett broke off the nearest twig, and held it before her prim face as she talked. She had a pretty childlike smile that came and went suddenly, but her face was not one that bore the marks of many pleasures. Mrs Flagg was a tall, commanding sort of person, with an air of satisfaction and authority.

'Oh, yes, gather all you want,' she said stiffly, as Miss Pickett took the syringa without having asked beforehand; but she had an amiable expression, and just now her large countenance was lighted up by pleasant anticipation.

'We can tell early what sort of a day it's goin' to be,' she said eagerly. 'There ain't a cloud in the sky now. I'll stop for you as I come along, or if there should be anything unforeseen to detain me, I'll send you word. I don't expect you'd want to go if it wa'n't so that I could?'

'Oh my sakes, no!' answered Miss Pickett discreetly, with a timid flush. 'You feel certain that Mis' Timms won't be put out? I shouldn't feel free to go unless I went 'long o' you.'

'Why, nothin' could be plainer than her words,' said Mrs Flagg in a tone of reproval. 'You saw how she urged me, an' had over all that talk about how we used to see each other often when we both lived to Longport, and told how she'd been thinkin' of writin', and askin' if it wa'n't so I should be able to come over and stop three or four days as soon as settled weather come, because she couldn't make no fire in her best chamber on account of the chimbley smokin' if the wind wa'n't just right. You see how she felt toward me, kissin' of me comin' and goin'? Why, she even asked me who I employed to do over my bonnet, Miss Pickett, just as interested as if she was a sister; an' she remarked she should look for us any pleasant day after we all got home, an' were settled after the conference.'

Miss Pickett smiled, but did not speak, as if she expected more arguments still.

'An' she seemed just about as much gratified to meet with you again. She seemed to desire to meet you again very particular,' continued Mrs Flagg. 'She really urged us to come together an' have a real good day talkin' over old times—there, don't le' 's go all over it again! I've always heard she'd made that old house of her aunt Bascoms' where she lives look real handsome. I once heard her best parlor carpet described as being an elegant carpet, different from any there was round here. Why, nobody couldn't be more cordial, Miss Pickett; you ain't goin' to give out just at the last?'

'Oh, no!' answered the visitor hastily; 'no, 'm! I want to go full as much as you do, Mis' Flagg, but you see I never was so well acquainted with Mis' Cap'n Timms, an' I always seem to dread putting myself for'ard. She certain was very urgent, an' she said plain enough to come any day next week, an' here 't is Wednesday, though of course she wouldn't look for us either Monday or Tuesday. 'T will be a real pleasant occasion, an' now we've been to the conference it don't seem near so much effort to start.'

'Why, I don't think nothin' of it,' said Mrs Flagg proudly. 'We shall have a grand good time, goin' together an' all, I feel sure.'

Miss Pickett still played with her syringa flower, tapping her thin cheek, and twirling the stem with her fingers. She looked as if she were going to say something more, but after a moment's hesitation she turned away.

'Good-afternoon, Mis' Flagg,' she said formally, looking up with a quick little smile; 'I enjoyed my call; I hope I ain't kep' you too late; I don't know but what it's 'most tea-time. Well, I shall look for you in the mornin'.'

'Good-afternoon, Miss Pickett; I'm glad I was in when you came. Call again, won't you?' said Mrs Flagg. 'Yes; you may expect me in good season,' and so they parted. Miss Pickett went out at the neat clicking gate in the white fence, and Mrs Flagg a moment later looked out of her sitting-room window to see if the gate were latched, and felt the least bit disappointed to find that it was. She sometimes went out after the departure of a guest, and fastened the gate herself with a loud, rebuking sound. Both of these Woodville women lived alone, and were very precise in their way of doing things.

II

THE next morning dawned clear and bright, and Miss Pickett rose even earlier than usual. She found it most difficult to decide which of her dresses would be best to wear. Summer was still so young that the day had all the freshness of spring, but when the two friends walked away together along the shady street, with a chorus of golden robins singing high overhead in the elms, Miss Pickett decided that she had made a wise choice of her second-best black silk gown, which she had just turned again* and freshened. It was neither too warm for the season nor too cool, nor did it look overdressed. She wore her large cameo pin, and this, with a long watch-chain, gave an air of proper mural decoration. She was a straight, flat little person, as if, when not in use, she kept herself, silk dress and all, between the leaves of a book. She carried a noticeable parasol with a fringe, and a small shawl, with a pretty border, neatly folded over her left arm. Mrs Flagg always dressed in black cashmere, and looked, to hasty observers, much the same one day as another; but her companion recognized the fact that this was the best black cashmere of all, and for a moment quailed at the thought that Mrs Flagg was paying such extreme deference to their prospective hostess. The visit turned for a moment into an unexpectedly solemn formality, and pleasure seemed to wane before Cynthia Pickett's eyes, yet with great courage she never slackened a single step. Mrs Flagg carried a somewhat worn black leather handbag, which Miss Pickett regretted; it did not give the visit that casual and unpremeditated air which she felt to be more elegant.

'Sha'n't I carry your bag for you?' she asked timidly. Mrs Flagg was the older and more important person.

'Oh, dear me, no,' answered Mrs Flagg. 'My pocket's so remote, in case I should desire to sneeze or anything, that I thought 't would be convenient for carrying my handkerchief and pocket-book; an' then I just tucked in a couple o' glasses o' my crab-apple jelly for Mis' Timms. She used to be a great hand for preserves of every sort, an' I thought 't would be a kind of an attention, an' give rise to conversation. I know she used to make excellent drop-cakes* when we was both residin' to Longport; folks used to say she never would give the right receipt, but if I get a real good chance, I mean to ask her. Or why can't you, if I start talkin' about receipts—why can't you say, sort of innocent, that I have always spoken frequently of her drop-cakes, an' ask for the rule? She would be very sensible to the compliment, and could pass it off if she didn't feel to indulge us. There, I do so wish you would!'

'Yes, 'm,' said Miss Pickett doubtfully; 'I'll try to make the opportunity. I'm very partial to drop-cakes. Was they flour or rye, Mis' Flagg?'

'They was flour, dear,' replied Mrs Flagg approvingly; 'crisp an' light as any you ever see.'

'I wish I had thought to carry somethin' to make it pleasant,' said Miss Pickett, after they had walked a little farther; 'but there, I don't know 's 't would look just right, this first visit, to offer anything to such a person as Mis' Timms. In case I ever go over to Baxter again I won't forget to make her some little present, as nice as I've got. 'T was certain very polite of her to urge me to come with you. I did feel very doubtful at first. I didn't know but she thought it behooved her, because I was in your company at the conference, and she wanted to save my feelin's, and yet expected I would decline. I never was well acquainted with her; our folks wasn't well off when I first knew her; 't was before uncle Cap'n Dyer passed away an' remembered mother an' me in his will. We couldn't make no han'some companies in them days, so we didn't go to none, an' kep' to ourselves; but in my grandmother's time, mother always said, the families was very friendly. I shouldn't feel like goin' over to pass the day with Mis' Timms if I didn't mean to ask her to return the visit. Some don't think o' these things, but mother was very set about not bein' done for when she couldn't make no return.'

'"When it rains porridge hold up your dish,"'* said Mrs Flagg; but Miss Pickett made no response beyond a feeble 'Yes, 'm,' which somehow got caught in her pale-green bonnet-strings.

'There, 't ain't no use to fuss too much over all them things,' proclaimed Mrs Flagg, walking along at a good pace with a fine sway of her skirts, and carrying her head high. 'Folks walks right by an' forgits all about you; folks can't always be going through with just so much. You'd had a good deal better time, you an' your ma, if you'd been freer in your ways; now don't you s'pose you would? 'T ain't what you give folks to eat so much as 't is makin' 'em feel welcome. Now, there's Mis' Timms; when we was to Longport she was dreadful methodical. She wouldn't let Cap'n Timms fetch nobody home to dinner without lettin' of her know, same 's other cap'ns' wives had to submit to. I was thinkin', when she was so cordial over to Danby, how she'd softened with time. Years do learn folks somethin'! She did seem very pleasant an' desirous. There, I am so glad we got started; if she'd gone an' got up a real good dinner to-day, an' then not had us come till to-morrow, 't would have been real too bad. Where anybody lives alone such a thing is very tryin'.'

'Oh, so 't is!' said Miss Pickett. 'There, I'd like to tell you what I went through with year before last. They come an' asked me one Saturday night to entertain the minister, that time we was having candidates'——

'I guess we'd better step along faster,' said Mrs Flagg suddenly. 'Why, Miss Pickett, there's the stage comin' now! It's dreadful prompt, seems to me. Quick! there's folks awaitin', an' I sha'n't get to Baxter in no state to visit Mis' Cap'n Timms if I have to ride all the way there backward!'

III

THE stage was not full inside. The group before the store proved to be made up of spectators, except one man, who climbed at once to a vacant seat by the driver. Inside there was only one person, after two passengers got out, and she preferred to sit with her back to the horses, so that Mrs Flagg and Miss Pickett settled themselves comfortably in the coveted corners of the back seat. At first they took no notice of their companion, and spoke to each other in low tones, but presently something attracted the attention of all three and engaged them in conversation.

'I never was over this road before,' said the stranger. 'I s'pose you ladies are well acquainted all along.'

'We have often traveled it in past years. We was over this part of it last week goin' and comin' from the county conference,' said Mrs Flagg is a dignified manner.

'What persuasion?' inquired the fellow-traveler, with interest.

'Orthodox,' said Miss Pickett quickly, before Mrs Flagg could speak. 'It was a very interestin' occasion; this other lady an' me stayed through all the meetin's.'

'I ain't Orthodox,' announced the stranger, waiving any interest in personalities. 'I was brought up amongst the Freewill Baptists.'*

'We're well acquainted with several of that denomination in our place,' said Mrs Flagg, not without an air of patronage. 'They've never built 'em no church; there ain't but a scattered few.'

'They prevail where I come from,' said the traveler. 'I'm goin' now to visit with a Freewill lady. We was to a conference together once, same 's you an' your friend, but 't was a state conference. She asked me to come some time an' make her a good visit, and I'm on my way now. I didn't seem to have nothin' to keep me to home.'

'We're all goin' visitin' to-day, ain't we?' said Mrs Flagg sociably; but no one carried on the conversation.

The day was growing very warm; there was dust in the sandy road, but the fields of grass and young growing crops looked fresh and fair. There was a light haze over the hills, and birds were thick in the air. When the stage-horses stopped to walk, you could hear the crows caw, and the bobolinks singing, in the meadows. All the farmers were busy in their fields.

'It don't seem but little ways to Baxter, does it?' said Miss Pickett, after a while. 'I felt we should pass a good deal o' time on the road, but we must be pretty near half-way there a'ready.'

'Why, more 'n half!' exclaimed Mrs Flagg. 'Yes; there's Beckett's Corner right ahead, an' the old Beckett house. I haven't been on this part of the road for so long that I feel kind of strange. I used to visit over here when I was a girl. There's a nephew's widow owns the place now. Old Miss Susan Beckett willed it to him, an' he died; but she resides there an' carries on the farm, an unusual smart woman, everybody says. Ain't it pleasant here, right out among the farms!'

'Mis' Beckett's place, did you observe?' said the stranger, leaning forward to listen to what her companions said. 'I expect that's where I'm goin'—Mis' Ezra Beckett's?'

'That's the one,' said Miss Pickett and Mrs Flagg together, and they both looked out eagerly as the coach drew up to the front door of a large old yellow house that stood close upon the green turf of the roadside.

The passenger looked pleased and eager, and made haste to leave the stage with her many bundles and bags. While she stood impatiently tapping at the brass knocker, the stage-driver landed a large trunk, and dragged it toward the door across the grass. Just then a busy-looking middle-aged woman made her appearance, with floury hands and a look as if she were prepared to be somewhat on the defensive.

'Why, how do you do, Mis' Beckett?' exclaimed the guest. 'Well, here I be at last. I didn't know 's you thought I was ever comin'. Why, I do declare, I believe you don't recognize me, Mis' Beckett.'

'I believe I don't,' said the self-possessed hostess. 'Ain't you made some mistake, ma'am?'

'Why, don't you recollect we was together that time to the state conference, an' you said you should be pleased to have me come an' make you a visit some time, an' I said I would certain. There, I expect I look more natural to you now.'

Mrs Beckett appeared to be making the best possible effort, and gave a bewildered glance, first at her unexpected visitor, and then at the trunk. The stage-driver, who watched this encounter with evident delight, turned away with reluctance. 'I can't wait all day to see how they settle it,' he said, and mounted briskly to the box, and the stage rolled on.

'He might have waited just a minute to see,' said Miss Pickett indignantly, but Mrs Flagg's head and shoulders were already far out of the stage window—the house was on her side. 'She ain't got in yet,' she told Miss Pickett triumphantly. 'I could see 'em quite a spell. With that trunk, too! I do declare, how inconsiderate some folks is!'

' 'T was pushin' an acquaintance most too far, wa'n't it?' agreed Miss Pickett. 'There, 't will be somethin' laughable to tell Mis' Timms. I never see anything more divertin'. I shall kind of pity that woman if we have to stop an' git her as we go back this afternoon.'

'Oh, don't let's forgit to watch for her,' exclaimed Mrs Flagg, beginning to brush off the dust of travel. 'There, I feel an excellent appetite, don't you? And we ain't got more 'n three or four miles to go, if we have that. I wonder what Mis' Timms is likely to give us for dinner; she spoke of makin' a good many chicken-pies, an' I happened to remark how partial I was to 'em. She felt above most of the things we had provided for us over to the conference. I know she was always counted the best o' cooks when I knew her so well to Long-port. Now, don't you forget, if there's a suitable opportunity, to inquire about the drop-cakes;' and Miss Pickett, a little less doubtful than before, renewed her promise.

IV

'MY gracious, won't Mis' Timms be pleased to see us! It's just exactly the day to have company. And ain't Baxter a sweet pretty place?' said Mrs Flagg, as they walked up the main street. 'Cynthy Pickett, now ain't you proper glad you come?

I felt sort o' calm about it part o' the time yesterday, but I ain't felt so like a girl for a good while. I do believe I'm goin' to have a splendid time.'

Miss Pickett glowed with equal pleasure as she paced along. She was less expansive and enthusiastic than her companion, but now that they were fairly in Baxter, she lent herself generously to the occasion. The social distinction of going away to spend a day in company with Mrs Flagg was by no means small. She arranged the folds of her shawl more carefully over her arm so as to show the pretty palm-leaf border, and then looked up with great approval to the row of great maples that shaded the broad sidewalk. 'I wonder if we can't contrive to make time to go an' see old Miss Nancy Fell?' she ventured to ask Mrs Flagg. 'There ain't a great deal o' time before the stage goes at four o'clock; 't will pass quickly, but I should hate to have her feel hurt. If she was one we had visited often at home, I shouldn't care so much, but such folks feel any little slight. She was a member of our church; I think a good deal of that.'

'Well, I hardly know what to say,' faltered Mrs Flagg coldly. 'We might just look in a minute; I shouldn't want her to feel hurt.'

'She was one that always did her part, too,' said Miss Pickett, more boldly. 'Mr Cronin used to say that she was more generous with her little than many was with their much.* If she hadn't lived in a poor part of the town, and so been occupied with a different kind of people from us, 't would have made a difference. They say she's got a comfortable little home over here, an' keeps house for a nephew. You know she was to our meeting one Sunday last winter, and 'peared dreadful glad to get back; folks seemed glad to see her, too. I don't know as you were out.'

'She always wore a friendly look,' said Mrs Flagg indulgently. 'There, now, there's Mis' Timms's residence; it's handsome, ain't it, with them big spruce-trees? I expect she may be at the window now, an' see us as we come along. Is my bonnet on straight, an' everything? The blinds looks open in the room this way; I guess she's to home fast enough.'

The friends quickened their steps, and with shining eyes

and beating hearts hastened forward. The slightest mists of uncertainty were now cleared away; they gazed at the house with deepest pleasure; the visit was about to begin.

They opened the front gate and went up the short walk, noticing the pretty herringbone pattern of the bricks, and as they stood on the high steps Cynthia Pickett wondered whether she ought not to have worn her best dress, even though there was lace at the neck and sleeves, and she usually kept it for the most formal of tea-parties and exceptional parish festivals. In her heart she commended Mrs Flagg for that familiarity with the ways of a wider social world which had led her to wear the very best among her black cashmeres.

'She's a good while coming to the door,' whispered Mrs Flagg presently. 'Either she didn't see us, or else she's slipped upstairs to make some change, an' is just goin' to let us ring again. I've done it myself sometimes. I'm glad we come right over after her urgin' us so; it seems more cordial than to keep her expectin' us. I expect she'll urge us terribly to remain with her over-night.'

'Oh, I ain't prepared,' began Miss Pickett, but she looked pleased. At that moment there was a slow withdrawal of the bolt inside, and a key was turned, the front door opened, and Mrs Timms stood before them with a smile. Nobody stopped to think at that moment what kind of smile it was.

'Why, if it ain't Mis' Flagg,' she exclaimed politely, 'an' Miss Pickett too! I am surprised!'

The front entry behind her looked well furnished, but not exactly hospitable; the stairs with their brass rods looked so clean and bright that it did not seem as if anybody had ever gone up or come down. A cat came purring out, but Mrs Timms pushed her back with a determined foot, and hastily closed the sitting-room door. Then Miss Pickett let Mrs Flagg precede her, as was becoming, and they went into a darkened parlor, and found their way to some chairs, and seated themselves solemnly.

''T is a beautiful day, ain't it?' said Mrs Flagg, speaking first. 'I don't know 's I ever enjoyed the ride more. We've been having a good deal of rain since we saw you at the conference, and the country looks beautiful.'

'Did you leave Woodville this morning? I thought I hadn't heard you was in town,' replied Mrs Timms formally. She was seated just a little too far away to make things seem exactly pleasant. The darkness of the best room seemed to retreat somewhat, and Miss Pickett looked over by the door, where there was a pale gleam from the side-lights in the hall, to try to see the pattern of the carpet; but her effort failed.

'Yes, 'm,' replied Mrs Flagg to the question. 'We left Woodville about half past eight, but it is quite a ways from where we live to where you take the stage. The stage does come slow, but you don't seem to mind it such a beautiful day.'

'Why, you must have come right to see me first!' said Mrs Timms, warming a little as the visit went on. 'I hope you're going to make some stop in town. I'm sure it was very polite of you to come right an' see me; well, it's very pleasant, I declare. I wish you'd been in Baxter last Sabbath; our minister did give us an elegant sermon on faith an' works. He spoke of the conference, and gave his views on some o' the questions that came up, at Friday evenin' meetin'; but I felt tired after getting home, an' so I wasn't out. We feel very much favored to have such a man amon'st us. He's building up the parish very considerable. I understand the pew-rents come to thirty-six dollars more this quarter than they did last.'

'We also feel grateful in Woodville for our pastor's efforts,' said Miss Pickett; but Mrs Timms turned her head away sharply, as if the speech had been untimely, and trembling Miss Pickett had interrupted.

'They're thinking here of raisin' Mr Barlow's salary another year,' the hostess added; 'a good many of the old parishioners have died off, but every one feels to do what they can. Is there much interest among the young people in Woodville, Mis' Flagg?'

'Considerable at this time, ma'am,' answered Mrs Flagg, without enthusiasm, and she listened with unusual silence to the subsequent fluent remarks of Mrs Timms.

The parlor seemed to be undergoing the slow processes of a winter dawn. After a while the three women could begin to see one another's faces, which aided them somewhat in carrying on a serious and impersonal conversation. There were a good many subjects to be touched upon, and Mrs Timms said everything that she should have said, except to invite her visitors to walk upstairs and take off their bonnets. Mrs Flagg sat her parlor-chair as if it were a throne, and carried her banner of self-possession as high as she knew how, but toward the end of the call even she began to feel hurried.

'Won't you ladies take a glass of wine an' a piece of cake after your ride?' inquired Mrs Timms, with an air of hospitality that almost concealed the fact that neither cake nor wine was anywhere to be seen; but the ladies bowed and declined with particular elegance. Altogether it was a visit of extreme propriety on both sides, and Mrs Timms was very pressing in her invitation that her guests should stay longer.

'Thank you, but we ought to be going,' answered Mrs Flagg, with a little show of ostentation, and looking over her shoulder to be sure that Miss Pickett had risen too. 'We've got some little ways to go,' she added with dignity. 'We should be pleased to have you call an' see us in case you have occasion to come to Woodville,' and Miss Pickett faintly seconded the invitation. It was in her heart to add, 'Come any day next week,' but her courage did not rise so high as to make the words audible. She looked as if she were ready to cry; her usual smile had burnt itself out into gray ashes; there was a white, appealing look about her mouth. As they emerged from the dim parlor and stood at the open front door, the bright June day, the golden-green trees, almost blinded their eyes. Mrs Timms was more smiling and cordial than ever.

'There, I ought to have thought to offer you fans; I am afraid you was warm after walking,' she exclaimed, as if to leave no stone of courtesy unturned. 'I have so enjoyed meeting you again, I wish it was so you could stop longer. Why, Mis' Flagg, we haven't said one word about old times when we lived to Longport. I've had news from there, too, since I saw you; my brother's daughter-in-law was here to pass the Sabbath after I returned.'

Mrs Flagg did not turn back to ask any questions as she stepped stiffly away down the brick walk. Miss Pickett followed her, raising the fringed parasol; they both made ceremonious little bows as they shut the high white gate behind them.

'Good-by,' said Mrs Timms finally, as she stood in the door with her set smile; and as they departed she came out and began to fasten up a rose-bush that climbed a narrow white ladder by the steps.

'Oh, my goodness alive!' exclaimed Mrs Flagg, after they had gone some distance in aggrieved silence, 'if I haven't gone and forgotten my bag! I ain't goin' back, whatever happens. I expect she'll trip over it in that dark room and break her neck!'

'I brought it; I noticed you'd forgotten it,' said Miss Pickett timidly, as if she hated to deprive her companion of even that slight consolation.

'There, I'll tell you what we'd better do,' said Mrs Flagg gallantly; 'we'll go right over an' see poor old Miss Nancy Fell; 't will please her about to death. We can say we felt like goin' somewhere to-day, an' 't was a good many years since either one of us had seen Baxter, so we come just for the ride, an' to make a few calls. She'll like to hear all about the conference; Miss Fell was always one that took a real interest in religious matters.'

Miss Pickett brightened, and they quickened their step. It was nearly twelve o'clock, they had breakfasted early, and now felt as if they had eaten nothing since they were grown up. An awful feeling of tiredness and uncertainty settled down upon their once buoyant spirits.

'I can forgive a person,' said Mrs Flagg, once, as if she were speaking to herself; 'I can forgive a person, but when I'm done with 'em, I'm done.'

V

'I DO declare, 't was like a scene in Scriptur' to see that poor good-hearted Nancy Fell run down her walk to open the gate for us!' said Mrs Persis Flagg later that afternoon, when she and Miss Pickett were going home in the stage. Miss Pickett nodded her head approvingly.

'I had a good sight better time with her than I should have had at the other place,' she said with fearless honesty. 'If I'd been Mis' Cap'n Timms, I'd made some apology or just passed us the compliment. If it wa'n't convenient, why couldn't she just tell us so after all her urgin' and sayin' how she should expect us?'

'I thought then she'd altered from what she used to be,' said Mrs Flagg. 'She seemed real sincere an' open away from home. If she wa'n't prepared to-day, 't was easy enough to say so; we was reasonable folks, an' should have gone away with none but friendly feelin's. We did have a grand good time with Nancy. She was as happy to see us as if we'd been queens.'

' 'T was a real nice little dinner,' said Miss Pickett gratefully. 'I thought I was goin' to faint away just before we got to the house, and I didn't know how I should hold out if she undertook to do anything extra, and keep us a-waitin'; but there, she just made us welcome, simple-hearted, to what she had. I never tasted such dandelion greens; an' that nice little piece o' pork and new biscuit, why, they was just splendid. She must have an excellent good cellar, if 't is such a small house. Her potatoes was truly remarkable for this time o' year. I myself don't deem it necessary to cook potatoes when I'm goin' to have dandelion greens. Now, didn't it put you in mind of that verse in the Bible that says, "Better is a dinner of herbs where love is"?* An' how desirous she'd been to see somebody that could tell her some particulars about the conference!'

'She'll enjoy tellin' folks about our comin' over to see her. Yes, I'm glad we went; 't will be of advantage every way, an' our bein' of the same church an' all, to Woodville. If Mis' Timms hears of our bein' there, she'll see we had reason, an' knew of a place to go. Well, I needn't have brought this old bag!'

Miss Pickett gave her companion a quick resentful glance, which was followed by one of triumph directed at the dust that was collecting on the shoulders of the best black cashmere; then she looked at the bag on the front seat, and suddenly felt illuminated with the suspicion that Mrs Flagg had secretly made preparations to pass the night in Baxter. The bag looked plump, as if it held much more than the pocket-book and the jelly.

Mrs Flagg looked up with unusual humility. 'I did think about that jelly,' she said, as if Miss Pickett had openly reproached her. 'I was afraid it might look as if I was tryin' to pay Nancy for her kindness.'

'Well, I don't know,' said Cynthia; 'I guess she'd been pleased. She'd thought you just brought her over a little present: but I do' know as 't would been any good to her after all; she'd thought so much of it, comin' from you, that she'd kep' it till 't was all candied.' But Mrs Flagg didn't look exactly pleased by this unexpected compliment, and her fellow-traveler colored with confusion and a sudden feeling that she had shown undue forwardness.

Presently they remembered the Beckett house, to their great relief, and, as they approached, Mrs Flagg reached over and moved her hand-bag from the front seat to make room for another passenger. But nobody came out to stop the stage, and they saw the unexpected guest sitting by one of the front windows comfortably swaying a palm-leaf fan, and rocking to and fro in calm content. They shrank back into their corners, and tried not to be seen. Mrs Flagg's face grew very red.

'She got in, didn't she?' said Miss Pickett, snipping her words angrily, as if her lips were scissors. Then she heard a call, and bent forward to see Mrs Beckett herself appear in the front doorway, very smiling and eager to stop the stage.

The driver was only too ready to stop his horses. 'Got a passenger for me to carry back, ain't ye?' said he facetiously. 'Them's the kind I like; carry both ways, make somethin' on a double trip,' and he gave Mrs Flagg and Miss Pickett a friendly wink as he stepped down over the wheel. Then he hurried toward the house, evidently in a hurry to put the baggage on; but the expected passenger still sat rocking and fanning at the window.

'No, sir; I ain't got any passengers,' exclaimed Mrs Beckett, advancing a step or two to meet him, and speaking very loud in her pleasant excitement. 'This lady that come this morning wants her large trunk with her summer things that she left to the depot in Woodville. She's very desirous to git into it, so don't you go an' forgit; ain't you got a book or somethin', Mr Ma'sh? Don't you forgit to make a note of it; here's her check, an' we've kep' the number in case you should mislay it or anything. There's things in the trunk she needs; you know how you overlooked stoppin' to the milliner's for my bunnit last week.'

'Other folks disremembers things as well 's me,' grumbled Mr Marsh. He turned to give the passengers another wink more familiar than the first, but they wore an offended air, and were looking the other way. The horses had backed a few steps, and the guest at the front window had ceased the steady motion of her fan to make them a handsome bow, and been puzzled at the lofty manner of their acknowledgment.

'Go 'long with your foolish jokes, John Ma'sh!' Mrs Beckett said cheerfully, as she turned away. She was a comfortable, hearty person, whose appearance adjusted the beauties of hospitality. The driver climbed to his seat, chuckling, and drove away with the dust flying after the wheels.

'Now, she's a friendly sort of a woman, that Mis' Beckett,' said Mrs Flagg unexpectedly, after a few moments of silence, when she and her friend had been unable to look at each other. 'I really ought to call over an' see her some o' these days, knowing her husband's folks as well as I used to, an' visitin' of 'em when I was a girl.' But Miss Pickett made no answer.

'I expect it was all for the best, that woman's comin',' suggested Mrs Flagg again hopefully. 'She looked like a will-ing person who would take right hold. I guess Mis' Beckett knows what she's about, and must have had her reasons. Perhaps she thought she'd chance it for a couple o' weeks anyway, after the lady'd come so fur, an' bein' one o' her own denomination. Hayin'-time'll be here before we know it. I think myself, gen'rally speakin', 't is just as well to let anybody know you're comin'.'

'Them seemed to be Mis' Cap'n Timms's views,' said Miss Pickett in a low tone; but the stage rattled a good deal, and Mrs Flagg looked up inquiringly, as if she had not heard.

A NEIGHBOR'S LANDMARK

I

THE timber-contractor took a long time to fasten his horse to the ring in the corner of the shed; but at last he looked up as if it were a matter of no importance to him that John Packer was coming across the yard. 'Good-day,' said he; 'good-day, John.' And John responded by an inexpressive nod.

'I was goin' right by, an' I thought I'd stop an' see if you want to do anything about them old pines o' yourn.'

'I don't know 's I do, Mr Ferris,' said John stiffly.

'Well, that business is easy finished,' said the contractor, with a careless air and a slight look of disappointment. 'Just as you say, sir. You was full of it a spell ago, and I kind o' kep' the matter in mind. It ain't no plot o' mine, 'cept to oblige you. I don't want to move my riggin' nowhere for the sake o' two trees—one tree, you might say; there ain't much o' anything but fire-wood in the sprangly* one. I shall end up over on the Foss lot next week, an' then I'm goin' right up country quick 's I can, before the snow begins to melt.'

John Packer's hands were both plunged deep into his side pockets, and the contractor did not fail to see that he was moving his fingers nervously.

'You don't want 'em blowin' down, breakin' all to pieces right on to your grass-land. They'd spile* pretty near an acre fallin' in some o' them spring gales. Them old trees is awful brittle. If you're ever calc'latin' to sell 'em, now's your time; the sprangly one's goin' back a'ready. They take the goodness all out o' that part o' your field, anyway,' said Ferris, casting a sly glance as he spoke.

'I don't know 's I care; I can maintain them two trees,' answered Packer, with spirit; but he turned and looked away, not at the contractor.

'Come, I mean business. I'll tell you what I'll do: if you want to trade, I'll give you seventy-five dollars for them two trees, and it's an awful price. Buyin' known trees like them's like tradin' for a tame calf; you'd let your forty-acre piece go without no fuss. Don't mind what folks say. They're yourn, John; or ain't they?'

'I'd just as soon be rid on 'em; they've got to come down some time,' said Packer, stung by this bold taunt. 'I ain't goin' to give you a present o' half their value, for all o' that.'

'You can't handle 'em yourself, nor nobody else about here; there ain't nobody got proper riggin' to handle them butts but me. I've got to take 'em down for ye fur 's I can see,' said Ferris, looking sly, and proceeding swiftly from persuasion to final arrangements. 'It's some like gittin' a tooth hauled; you kind o' dread it, but when 't is done you feel like a man. I ain't said nothin' to nobody, but I hoped you'd do what you was a-mind to with your own property. You can't afford to let all that money rot away; folks won't thank ye.'

'What you goin' to give for 'em?' asked John Packer impatiently. 'Come, I can't talk all day.'

'I'm a-goin' to give you seventy-five dollars in bank-bills,' said the other man, with an air of great spirit.

'I ain't a-goin' to take it, if you be,' said John, turning round, and taking a hasty step or two toward the house. As he turned he saw the anxious faces of two women at one of the kitchen windows, and the blood flew to his pinched face.

'Here, come back here and talk man-fashion!' shouted the timber-dealer. 'You couldn't make no more fuss if I come to seize your farm. I'll make it eighty, an' I'll tell you jest one thing more: if you're holdin' out, thinkin' I'll give you more, you can hold out till doomsday.'

'When'll you be over?' said the farmer abruptly; his hands were clenched now in his pockets. The two men stood a little way apart, facing eastward, and away from the house. The long, wintry fields before them sloped down to a wide stretch of marshes covered with ice, and dotted here and there with an abandoned haycock. Beyond was the gray sea less than a mile away; the far horizon was like an edge of steel. There was a small fishing-boat standing in toward the shore, and far off were two or three coasters.

'Looks cold, don't it?' said the contractor. 'I'll be over middle o' the week some time, Mr Packer.' He unfastened his horse, while John Packer went to the unsheltered wood-pile and began to chop hard at some sour, heavy-looking pieces of red-oak wood. He stole a look at the window, but the two troubled faces had disappeared.

II

LATER that afternoon John Packer came in from the barn; he had lingered out of doors in the cold as long as there was any excuse for so doing, and had fed the cattle early, and cleared up and laid into a neat pile some fencing materials and pieces of old boards that had been lying in the shed in great confusion since before the coming of snow. It was a dusty, splintery heap, half worthless, and he had thrown some of the broken fence-boards out to the wood-pile, and then had stopped to break them up for kindlings and to bring them into the back kitchen of the house, hoping, yet fearing at every turn, to hear the sound of his wife's voice. Sometimes the women had to bring in fire-wood themselves, but to-night he filled the great wood-box just outside the kitchen door, piling it high with green beech and maple, with plenty of dry birch and pine, taking pains to select the best and straightest sticks, even if he burrowed deep into the wood-pile. He brought the bushel basketful of kindlings last, and set it down with a cheerful grunt, having worked himself into good humor again; and as he opened the kitchen door, and went to hang his great blue mittens behind the stove, he wore a self-satisfied and pacificatory smile.

'There, I don't want to hear no more about the wood-box bein' empty. We're goin' to have a cold night; the air's full of snow, but 't won't fall, not till it moderates.'

The women glanced at him with a sense of relief. They had looked forward to his entrance in a not unfamiliar mood of surly silence. Every time he had thumped down a great armful of wood, it had startled them afresh, and their timid protest and sense of apprehension had increased until they were pale and miserable; the younger woman had been crying.

'Come, mother, what you goin' to get me for supper?'

said the master of the house. 'I'm goin' over to the Centre to the selec'men's office to-night. They're goin' to have a hearin' about that new piece o' road over in the Dexter neighborhood.'

The mother and daughter looked at each other with relief and shame; perhaps they had mistaken the timber-contractor's errand, after all, though their imagination had followed truthfully every step of a bitter bargain, from the windows.

'Poor father!' said his wife, half unconsciously. 'Yes; I'll get you your supper quick 's I can. I forgot about to-night. You'll want somethin' warm before you ride 'way over to the Centre, certain;' and she began to bustle about, and to bring things out of the pantry. She and John Packer had really loved each other when they were young, and although he had done everything he could since then that might have made her forget, she always remembered instead; she was always ready to blame herself, and to find excuse for him. 'Do put on your big fur coat, won't you, John?' she begged eagerly.

'I ain't gone yet,' said John, looking again at his daughter, who did not look at him. It was not quite dark, and she was bending over her sewing, close to the window. The momentary gleam of hope had faded in her heart; her father was too pleasant: she hated him for the petty deceit.

'What are you about there, Lizzie?' he asked gayly. 'Why don't you wait till you have a light? Get one for your mother: she can't see over there by the table.'

Lizzie Packer's ready ears caught a provoking tone in her father's voice, but she dropped her sewing, and went to get the hand-lamp from the high mantelpiece. 'Have you got a match in your pocket? You know we're all out; I found the last this mornin' in the best room.' She stood close beside him while he took a match from his waistcoat pocket and gave it to her.

'I won't have you leavin' matches layin' all about the house,' he commanded; 'mice'll get at 'em, and set us afire. You can make up some lamplighters out of old letters and things; there's a lot o' stuff that might be used up. Seems to me lamplighters is gone out o' fashion; they come in very handy.'

Lizzie did not answer, which was a disappointment.

'Here, you take these I've got in my pocket, and that'll remind me to buy some at the store,' he ended. But Lizzie did not come to take them, and when she had waited a moment, and turned up the lamp carefully, she put it on the table by her mother, and went out of the room. The father and mother heard her going upstairs.

'I do hope she won't stay up there in the cold,' said Mrs Packer in an outburst of anxiety.

'What's she sulkin' about now?' demanded the father, tipping his chair down emphatically on all four legs. The timid woman mustered all her bravery.

'Why, when we saw Mr Ferris out there talkin' with you, we were frightened for fear he was tryin' to persuade you about the big pines. Poor Lizzie got all worked up; she took on and cried like a baby when we saw him go off chucklin' and you stayed out so long. She can't bear the thought o' touchin' 'em. And then when you come in and spoke about the selec'men, we guessed we was all wrong. Perhaps Lizzie feels bad about that now. I own I had hard feelin's toward you myself, John.' She came toward him with her mixing-spoon in her hand; her face was lovely and hopeful. 'You see, they've been such landmarks, John,' she said, 'and our Lizzie's got more feelin' about 'em than anybody. She was always playin' around 'em when she was little; and now there's so much talk about the fishin' folks countin' on 'em to get in by the short channel in bad weather, and she don't want you blamed.'

'You'd ought to set her to work, and learnt her head to save her heels,' said John Packer, grumbling; and the pale little woman gave a heavy sigh, and went back to her work again. 'That's why she ain't no good now—playin' out all the time when other girls was made to work. Broke you all down, savin' her,' he ended in an aggrieved tone.

'John, 't ain't true, is it?' She faced him again in a way that made him quail; his wife was never disrespectful, but she sometimes faced every danger to save him from his own foolishness. 'Don't you go and do a thing to make everybody hate you. You know what it says in the Bible about movin' a landmark.* You'll get your rights; 't is just as much your right to let the trees stand, and please folks.'

'Come, come, Mary Hannah!' said John, a little moved in spite of himself. 'Don't work yourself up so. I ain't told you I was goin' to cut 'em, have I? But if I ever do, 't is because I've been twitted into it, an' told they were everybody's trees but mine.'

He pleased himself at the moment by thinking that he could take back his promise to Ferris, even if it cost five dollars to do it. Why couldn't people leave a man alone? It was the women's faces at the window that had decided his angry mind, but now they thought it all his fault. Ferris would say, 'So your women folks persuaded you out of it.' It would be no harm to give Ferris a lesson: he had used a man's being excited and worked upon by interfering neighbors to drive a smart bargain. The trees were worth fifty dollars apiece, if they were worth a cent. John Packer transferred his aggrieved thoughts from his family to Ferris himself. Ferris had driven a great many sharp bargains; he had plenty of capital behind him, and had taken advantage of the hard times, and of more than one man's distress, to buy woodland at far less than its value. More than that, he always stripped land to the bare skin; if the very huckleberry bushes and ferns had been worth anything to him, he would have taken those, insisting upon all or nothing, and, regardless of the rights of forestry, he left nothing to grow; no sapling-oak or pine stood where his hand had been. The pieces of young growing woodland that might have made their owners rich at some later day were sacrificed to his greed of gain. You had to give him half your trees to make him give half price for the rest. Some men yielded to him out of ignorance, or avarice for immediate gains, and others out of bitter necessity. Once or twice he had even brought men to their knees and gained his point by involving them in money difficulties, through buying up their mortgages and notes. He could sell all the wood and timber he could buy, and buy so cheap, to larger dealers; and a certain builder having given him an order for some unusually wide and clear pine at a large price, his withering eye had been directed toward the landmark trees on John Packer's farm.

On the road home from the Packer farm that winter afternoon Mr Ferris's sleighbells sounded lonely, and nobody was

met or overtaken to whom he could brag of his success. Now
and then he looked back with joy to the hill behind the
Packer house, where the assailed pine-trees still stood
together, superb survivors of an earlier growth. The snow was
white about them now, but in summer they stood near the
road at the top of a broad field which had been won from wild
land by generation after generation of the Packers. Whatever
man's hands have handled, and his thoughts have centred in,
gives something back to man, and becomes charged with his
transferred life, and brought into relationship. The great
pines could remember all the Packers, if they could remember anything; they were like some huge archaic creatures
whose thoughts were slow and dim. So many anxious eyes had
sought these trees from the sea, so many wanderers by land
had gladly welcomed the far sight of them in coming back to
the old town, it must have been that the great live things felt
their responsibility as landmarks and sentinels. How could
any fisherman find the deep-sea fishing-grounds for cod and
haddock without bringing them into range with a certain blue
hill far inland, or with the steeple of the old church on the
Wilton road? How could a hurrying boat find the short way
into harbor before a gale without sighting the big trees from
point to point among the rocky shallows? It was a dangerous
bit of coast in every way, and every fisherman and pleasure-
boatman knew the pines on Packer's Hill. As for the Packers
themselves, the first great adventure for a child was to climb
alone to the great pines, and to see an astonishing world from
beneath their shadow; and as the men and women of the
family grew old, they sometimes made an effort to climb
the hill once more in summer weather, to sit in the shelter of
the trees, where the breeze was cool, and to think of what
had passed, and to touch the rough bark with affectionate
hands. The boys went there when they came home from
voyages at sea; the girls went there with their lovers. The
trees were like friends, and whether you looked seaward,
being in an inland country, or whether you looked shore-
ward, being on the sea, there they stood and grew in their
places, while a worldful of people lived and died, and again
and again new worldfuls were born and passed away, and still

these landmark pines lived their long lives, and were green and vigorous yet.

III

THERE was a fishing-boat coming into the neighboring cove, as has already been said, while Ferris and John Packer stood together talking in the yard. In this fishing-boat were two other men, younger and lighter-hearted, if it were only for the reason that neither of them had such a store of petty ill deeds and unkindnesses to remember in dark moments. They were in an old dory, and there was much ice clinging to her, inside and out, as if the fishers had been out for many hours. There were only a few cod lying around in the bottom, already stiffened in the icy air. The wind was light, and one of the men was rowing with short, jerky strokes, to help the sail, while the other held the sheet and steered with a spare oar that had lost most of its blade. The wind came in flaws, chilling, and mischievous in its freaks. 'I ain't goin' out any more this year,' said the younger man, who rowed, giving a great shudder. 'I ain't goin' to perish myself for a pinch o' fish like this'—pushing them with his heavy boot. 'Generally it's some warmer than we are gittin' it now, 'way into January. I've got a good chance to go into Otis's shoe-shop; Bill Otis was tellin' me he didn't know but he should go out West to see his uncle's folks,—he done well this last season, lobsterin',—an' I can have his bench if I want it. I do' know but I may make up some lobster-pots myself, evenin's an' odd times, and take to lobsterin' another season. I know a few good places that Bill Otis ain't struck; and then the scarcer lobsters git to be, the more you git for 'em, so now a poor ketch's 'most better 'n a good one.'

'Le' me take the oars,' said Joe Banks, without attempting a reply to such deep economical wisdom.

'You hold that sheet light,' grumbled the other man, 'or these gusts'll have us over. An' don't let that old oar o' yourn range about so. I can't git no hold o' the water.' The boat lifted suddenly on a wave and sank again in the trough, the sail flapped, and a great cold splash of salt water came aboard, floating the fish to the stern, against Banks's feet. Chauncey, grumbling heartily, began to bail with a square-built wooden scoop for which he reached far behind him in the bow.

'They say the sea holds its heat longer than the land, but I guess summer's about over out here.' He shivered again as he spoke. 'Come, le' 's say this is the last trip, Joe.'

Joe looked up at the sky, quite unconcerned. 'We may have it warmer after we git more snow,' he said. 'I'd like to keep on myself until after the first o' the year, same 's usual. I've got my reasons,' he added. 'But don't you go out no more, Chauncey.'

'What you goin' to do about them trees o' Packer's?' asked Chauncey suddenly, and not without effort. The question had been on his mind all the afternoon. 'Old Ferris had laid a bet that he'll git 'em anyway. I signed the paper they've got down to Fox'l Berry's store to the Cove. A number has signed it, but I shouldn't want to be the one to carry it up to Packer. They all want your name, but they've got some feelin' about how you're situated. Some o' the boys made me promise to speak to you, bein' 's we're keepin' together.'

'You can tell 'em I'll sign it,' said Joe Banks, flushing a warm, bright color under his sea-chilled skin. 'I don't know what set him out to be so poor-behaved. He's a quick-tempered man, Packer is, but quick over. I never knew him to keep no such a black temper as this.'

'They always say that you can't drive a Packer,' said Chauncey, tugging against the uneven waves. 'His mother came o' that old fightin' stock up to Bolton; 't was a different streak from his father's folks—they was different-hearted an' all pleasant. Ferris has done the whole mean business. John Packer'd be madder 'n he is now if he knowed how Ferris is makin' a tool of him. He got a little too much aboard long ago 's Thanksgivin' Day, and bragged to me an' another fellow when he was balmy how he'd rile up Packer into sellin' them pines, and then he'd double his money on 'em up to Boston; he said there wa'n't no such a timber pine as that big one left in the State that he knows on. Why, 't is 'most five foot through high 's I can reach.'

Chauncey stopped rowing a minute, and held the oars with one hand while he looked over his shoulder. 'I should miss them old trees,' he said; 'they always make me think of a married couple. They ain't no common growth, be they, Joe? Everybody knows 'em. I bet you if anything happened to one on 'em t' other would go an' die. They say ellums* has mates, an' all them big trees.'

Joe Banks had been looking at the pines all the way in; he had steered by them from point to point. Now he saw them just over Fish Rock, where the surf was whitening, and over the group of fish-houses, and began to steer straight inshore. The sea was less rough now, and after getting well into the shelter of the land he drew in his oar. Chauncey could pull the rest of the way without it. A sudden change in the wind filled the three-cornered sail, and they moved faster.

'She'll make it now, herself, if you'll just keep her straight, Chauncey; no, 't wa'n't nothin' but a flaw, was it? Guess I'd better help ye;' and he leaned on the oar once more, and took a steady sight of the familiar harbor marks.

'We're right over one o' my best lobster rocks,' said Chauncey, looking warm-blooded and cheerful again. 'I'm satisfied not to be no further out; it's beginnin' to snow; see them big flakes a-comin'? I'll tell the boys about your signin' the paper; I do' know 's you'd better resk it, either.'

'Why not?' said Joe Banks hastily. 'I suppose you refer to me an' Lizzie Packer; but she wouldn't think no more o' me for leavin' my name off a proper neighborhood paper, nor her father, neither. You git them two pines let alone, and I'll take care o' Lizzie. I've got all the other boats and men to think of besides me, an' I've got some pride anyway. I ain't goin' to have Bolton folks an' all on 'em down to the Centre twittin' us, nor twittin' Packer; he'll turn sour toward everybody the minute he does it. I know Packer; he's rough and ugly, but he ain't the worst man in town by a good sight. Anybody'd be all worked up to go through so much talk, and I'm kind o' 'fraid this minute his word's passed to Ferris to have them trees down. But you show him the petition; 't will be kind of formal, and if that don't do no good, I do' know what will. There you git the sail in while I hold her stiddy, Chauncey.'

IV

AFTER a day or two of snow that turned to rain, and was followed by warmer weather, there came one of the respites which keep up New England hearts in December. The short, dark days seemed shorter and darker than usual that year, but one morning the sky had a look of Indian summer, the wind was in the south, and the cocks and hens of the Packer farm came boldly out into the sunshine, to crow and cackle before the barn. It was Friday morning, and the next day was the day before Christmas.

John Packer was always good-tempered when the wind was in the south. The milder air, which relaxed too much the dispositions of less energetic men, and made them depressed and worthless, only softened and tempered him into reasonableness. As he and his wife and daughter sat at breakfast, after he had returned from feeding the cattle and horses, he wore a pleasant look, and finally leaned back and said the warm weather made him feel boyish, and he believed that he would take the boat and go out fishing.

'I can haul her out and fix her up for winter when I git ashore,' he explained. 'I've been distressed to think it wa'n't done before. I expect she's got some little ice in her now, there where she lays just under the edge of Joe Banks's fish-house. I spoke to Joe, but he said she'd do till I could git down. No; I'll turn her over, and make her snug for winter, and git a small boat o' Joe. I ain't goin' out a great ways: just so 's I can git a cod or two. I always begin to think of a piece o' new fish quick 's these mild days come; feels like the Janooary thaw.'

' 'T would be a good day for you to ride over to Bolton, too,' said Mrs Packer. 'But I'd like to go with you when you go there, an' I've got business here to-day. I've put the kettle on some time ago to do a little colorin'. We can go to Bolton some day next week.'

'I've got to be here next week,' said Packer ostentatiously; but at this moment his heart for the first time completely failed him about the agreement with Ferris. The south wind had blown round the vane of his determination. He forgot his wife and daughter, laid down his knife and fork, and quite unknown to himself began to hang his head. The great trees were not so far from the house that he had not noticed the sound of the southerly breeze in their branches as he came across the yard. He knew it as well as he knew the rote* of the beaches and ledges on that stretch of shore. He was meaning, at any rate, to think it over while he was out fishing, where nobody could bother him. He wasn't going to be hindered by a pack of folks from doing what he liked with his own; but neither was old Ferris going to say what he had better do with his own trees.

'You put me up a bite o' somethin' hearty, mother,' he made haste to say. 'I sha'n't git in till along in the afternoon.'

'Ain't you feelin' all right, father?' asked Lizzie, looking at him curiously.

'I be,' said John Packer, growing stern again for the moment. 'I feel like a day out fishin'. I hope Joe won't git the start o' me. You seen his small boat go out?' He looked up at his daughter, and smiled in a friendly way, and went on with his breakfast. It was evidently one of his pleasant days; he never had made such a frank acknowledgment of the lovers' rights, but he had always liked Joe Banks. Lizzie's cheeks glowed; she gave her mother a happy glance of satisfaction, and looked as bright as a rose. The hard-worked little woman smiled back in sympathy. There was a piece of her best loaf cake in the round wooden luncheon-box that day, and every-thing else that she thought her man would like and that his box would hold, but it seemed meagre to her generous heart even then. The two women affectionately watched him away down the field-path that led to the cove where the fish-houses were.

All the Wilton farmers near the sea took a turn now and then at fishing. They owned boats together sometimes, but John Packer had always kept a good boat of his own. To-day he had no real desire to find a companion or to call for help to launch his craft, but finding that Joe Banks was busy in his fish-house, he went in to borrow the light dory and a pair of oars. Joe seemed singularly unfriendly in his manner, a little cold and strange, and went on with his work without looking

up. Mr Packer made a great effort to be pleasant; the south
wind gave him even a sense of guilt.

'Don't you want to come, Joe?' he said, according to 'longshore etiquette; but Joe shook his head, and showed no
interest whatever. It seemed then as if it would be such a good
chance to talk over the tree business with Joe, and to make
him understand there had been some reason in it; but John
Packer could mind his own business as well as any man, and so
he picked his way over the slippery stones, pushed off the
dory, stepped in, and was presently well outside on his way to
Fish Rock. He had forgotten to look for any bait until Joe had
pushed a measure of clams along the bench; he remembered
it now as he baited his cod-lines, sitting in the swaying and
lifting boat, a mile or two out from shore. He had but poor
luck; the cold had driven the fish into deeper water, and
presently he took the oars to go farther out, and looking at
the land for the first time with a consciousness of seeing it, he
sighted his range, and turned the boat's head. He was still so
near land that beyond the marshes, which looked narrow
from the sea, he could see his own farm and his neighbors'
farms on the hill that sloped gently down; the northern point
of higher land that sheltered the cove and the fish-houses also
kept the fury of the sea winds from these farms, which faced
the east and south. The main road came along the high ridge
at their upper edge, and a lane turned off down to the cove;
you could see this road for three or four miles when you were
as far out at sea. The whole piece of country most familiar to
John Packer lay there spread out before him in the morning
sunshine. The house and barn and corn-house looked like
children's playthings; he made a vow that he would get out
the lumber that winter for a wood-shed; he needed another
building, and his wood-pile ought to be under cover. His wife
had always begged him to build a shed; it was hard for a
woman to manage with wet wood in stormy weather; often he
was away, and they never kept a boy or man to help with farm-
work except in summer. 'Joe Banks was terribly surly about
something,' said Mr Packer to himself. But Joe wanted Lizzie.
When they were married he meant to put an addition to the
farther side of the house, and to give Joe a chance to come

right there. Lizzie's mother was liable to be ailing, and needed her at hand. That eighty dollars would come in handy these hard times.

John Packer like to be cross and autocratic, and to oppose people; but there was hidden somewhere in his heart a warm spot of affectionateness and desire for approval. When he had quarreled for a certain time, he turned square about on this instinct as on a pivot. The self-love that made him wish to rule ended in making him wish to please; he could not very well bear being disliked. The bully is always a coward, but there was a good sound spot of right-mindedness, after all, in John Packer's gnarly disposition.

As the thought of the price of his trees flitted through John Packer's mind, it made him ashamed instead of pleasing him. He rowed harder for some distance, and then stopped to loosen the comforter about his neck. He looked back at the two pines where they stood black and solemn on the distant ridge against the sky. From this point of view they seemed to have taken a step nearer each other, as if each held the other fast with its branches in a desperate alliance. The bare, strong stem of one, the drooping boughs of the other, were indistinguishable, but the trees had a look as if they were in trouble. Something made John Packer feel sick and dizzy, and blurred his eyes so that he could not see them plain; the wind had weakened his eyes, and he rubbed them with his rough sleeve. A horror crept over him before he understood the reason, but in another moment his brain knew what his eyes had read. Along the ridge road came something that trailed long and black like a funeral, and he sprang to his feet in the dory, and lost his footing, then caught at the gunwale, and sat down again in despair. It was like the panic of a madman, and he cursed and swore at old Ferris for his sins, with nothing to hear him but the busy waves that glistened between him and the shore. Ferris had stolen his chance; he was coming along with his rigging as fast as he could, with his quick French wood-choppers, and their sharp saws and stubborn wedges to cant the trunks; already he was not far from the farm. Old Ferris was going to set up his yellow sawdust-mill there—that was the plan; the great trunks were too heavy to handle or haul any distance with any trucks or sleds that were used nowadays. It would be all over before anybody could get ashore to stop them; he would risk old Ferris for that.

Packer began to row with all his might; he had left the sail ashore. The oars grew hot at the wooden thole-pins, and he pulled and pulled. There would be three quarters of a mile to run up-hill to the house, and another bit to the trees themselves, after he got in. By that time the two-man saw, and the wedges, and the Frenchmen's shining axes, might have spoiled the landmark pines. 'Lizzie's there—she'll hold 'em back till I come,' he gasped, as he passed Fish Rock. 'Oh, Lord! what a fool! I ain't goin' to have them trees murdered;' and he set his teeth hard, and rowed with all his might.

Joe Banks looked out of the little four-paned fish-house window, and saw the dory coming, and hurried to the door. 'What's he puttin' in so for?' said he to himself, and looked up the coast to see if anything had happened; the house might be on fire. But all the quiet farms looked untroubled. 'He's pullin' at them oars as if the devil was after him,' said Joe to himself. 'He couldn't ha' heard o' that petition they're gettin' up from none of the fish he's hauled in; 't will 'bout set him crazy, but I was bound I'd sign it with the rest. The old dory's jumpin' right out of water every stroke he pulls.'

V

THE next night the Packer farmhouse stood in the winter landscape under the full moon, just as it had stood always, with a light in the kitchen window, and a plume of smoke above the great, square chimney. It was about half past seven o'clock. A group of men were lurking at the back of the barn, like robbers, and speaking in low tones. Now and then the horse stamped in the barn, or a cow lowed; a dog was barking, away over on the next farm, with an anxious tone, as if something were happening that he could not understand. The sea boomed along the shore beyond the marshes; the men could hear the rote of a piece of pebble beach a mile or two to the southward; now and then there was a faint tinkle of sleigh-bells. The fields looked wide and empty; the unusual warmth of the day before had been followed by clear cold. Suddenly a straggling company of women were seen coming from the next house. The men at the barn flapped their arms, and one of them, the youngest, danced a little to keep himself warm.

'Here they all come,' said somebody, and at that instant the sound of many sleighbells grew loud and incessant, and far-away shouts and laughter came along the wind, fainter in the hollows and loud on the hills of the uneven road. 'Here they come! I guess you'd better go in, Joe; they'll want to have lights ready.'

'She'll have a fire all laid for him in the fore room,' said the young man; 'that's all we want. She'll be expectin' you, Joe; go in now, and they'll think nothin' of it, bein' Saturday night. Just you hurry, so they'll have time to light up.' And Joe went.

'Stop and have some talk with father,' whispered Lizzie affectionately to her lover, as she came to meet him. 'He's all worked up, thinking nobody'll respect him, an' all that. Tell him you're glad he beat.' And they opened the kitchen door.

'What's all that noise?' said John Packer, dropping his weekly newspaper, and springing out of his chair. He looked paler and thinner than he had looked the day before. 'What's all that noise, Joe?'

There was a loud sound of bells now, and of people cheer-ing. Joe's throat had a lump in it; he knew well enough what it was, and could not find his voice to tell. Everybody in the neighborhood was coming, and they were all cheering as they passed the landmark pines.

'I guess the neighbors mean to give you a little party to-night, sir,' said Joe. 'I see six or eight sleighs comin' along the road. They've all heard about it; some o' the boys that was here with the riggin' went down to the store last night, and they was all tellin' how you stood right up to Ferris like a king, an' drove him. You see, they're all gratified on account of having you put a stop to Ferris's tricks about them pines,' he repeated. Joe did not dare to look at Lizzie or her mother, and in two minutes more the room began to fill with people,

and John Packer, who usually hated company, was shaking
hands hospitably with everybody that came.

Half an hour afterward, Mr Packer and Joe Banks and Joe's
friend Chauncey were down cellar together, filling some
pitchers from the best barrel of cider. The guests were tramping to and fro overhead in the best room; there was a great
noise of buzzing talk and laughter.

'Come, sir, give us a taste before we go up; it's master hot
up there,' said Chauncey, who was nothing if not convivial;
and the three men drank solemnly in turn from the smallest
of the four pitchers; then Mr Packer stooped again to replenish it.

'Whatever become o' that petition?' whispered Chauncey;
but Joe Banks gave him a warning push with his elbow. 'Wish
ye merry Christmas!' said Chauncey unexpectedly to some
one who called him from the stairhead.

'Hold that light nearer,' said Mr Packer. 'Come, Joe, I ain't
goin' to hear no more o' that nonsense about me beatin' off
old Ferris.' He had been king of his Christmas company
upstairs, but down here he was a little ashamed.

'Land! there's the fiddle,' said Chauncey. 'Le' 's hurry up;'
and the three cup-bearers hastened back up the cellar-stairs
to the scene of festivity.

The two Christmas trees, the landmark pines, stood tall and
strong on the hill looking down at the shining windows of the
house. There was a sound like a summer wind in their tops;
the bright moon and the stars were lighting them, and all the
land and sea, that Christmas night.

THE LIFE OF NANCY

I

THE wooded hills and pastures of eastern Massachusetts are so close to Boston that from upper windows of the city, looking westward, you can see the tops of pine-trees and orchard-boughs on the high horizon. There is a rustic environment on the landward side; there are old farmhouses at the back of Milton Hill and beyond Belmont which look as unchanged by the besieging suburbs of a great city as if they were forty miles from even its borders. Now and then, in Boston streets, you can see an old farmer in his sleigh or farm wagon as if you saw him in a Berkshire village. He seems neither to look up at the towers nor down at any fashionable citizens, but goes his way alike unconscious of seeing or being seen.

On a certain day a man came driving along Beacon Street, who looked bent in the shoulders, as if his worn fur cap were too heavy for head and shoulders both. This type of the ancient New England farmer in winter twitched the reins occasionally, like an old woman, to urge the steady white horse that plodded along as unmindful of his master's suggestions as of the silver-mounted harnesses that passed them by. Both horse and driver appeared to be conscious of sufficient wisdom, and even worth, for the duties of life; but all this placidity and self-assurance were in sharp contrast to the eager excitement of a pretty, red-cheeked girl who sat at the driver's side. She was as sensitive to every new impression as they were dull. Her face bloomed out of a round white hood in such charming fashion that those who began to smile at an out-of-date equipage were interrupted by a second and stronger instinct, and paid the homage that one must always pay to beauty.

It was a bitter cold morning. The great sleighbells on the horse's shaggy neck jangled along the street, and seemed to still themselves as they came among the group of vehicles that were climbing the long hill by the Common.*

As the sleigh passed a clubhouse that stands high on the slope, a young man who stood idly behind one of the large windows made a hurried step forward, and his sober face relaxed into a broad, delighted smile; then he turned quickly, and presently appearing at the outer door, scurried down the long flight of steps to the street, fastening the top buttons of his overcoat by the way. The old sleigh, with its worn buffalo skin hanging unevenly over the back, was only a short distance up the street, but its pursuer found trouble in gaining much upon the steady gait of the white horse. He ran two or three steps now and then, and was almost close enough to speak as he drew near to the pavement by the State House. The pretty girl was looking up with wonder and delight, but in another moment they went briskly on, and it was not until a long pause had to be made at the blocked crossing of Tremont Street that the chase was ended.

The wonders of a first visit to Boston were happily continued to Miss Nancy Gale in the sudden appearance at her side of a handsome young gentleman. She put out a most cordial and warm hand from her fitch muff, and her acquaintance noticed with pleasure the white knitted mitten that protected it from the weather. He had not yet found time to miss the gloves left behind at the club, but the warm little mitten was very comfortable to his fingers.

'I was just thinking—I hoped I should see you, when I was starting to come in this morning,' she said, with an eager look of pleasure; then, growing shy after the unconscious joy of the first moment, 'Boston is a pretty big place, isn't it?'

'We all think so,' said Tom Aldis with fine candor. 'It seems odd to see you here.'

'Uncle Ezra, this is Mr Aldis that I have been telling you about, who was down at our place so long in the fall,' explained Nancy, turning to look appealingly at her stern companion. 'Mr Aldis had to remain with a friend who had sprained his ankle. Is Mr Carew quite well now?' she turned again to ask.

'Oh yes,' answered Tom. 'I saw him last week; he's in New York this winter. But where are you staying, Nancy?' he asked eagerly, with a hopeful glance at uncle Ezra. 'I should like to take you somewhere this afternoon. This is your first visit, isn't it? Couldn't you go to see Rip Van Winkle to-morrow? It's the very best thing there is just now. Jefferson's playing this week.'*

'Our folks ain't in the habit of attending theatres, sir,' said uncle Ezra, checking this innocent plan as effectually as an untracked horse-car* was stopping traffic in the narrow street. He looked over his shoulder to see if there were any room to turn, but was disappointed.

Tom Aldis gave a glance, also, and was happily reassured; the street was getting fuller behind them every moment. 'I beg you to excuse me, sir,' he said gallantly to the old man. 'Do you think of anything else that Miss Gale ought to see? There is the Art Museum, if she hasn't been there already; all the pictures and statues and Egyptian things, you know.'

There was much deference and courtesy in the young man's behavior to his senior. Uncle Ezra responded by a less suspicious look at him, but seemed to be considering this new proposition before he spoke. Uncle Ezra was evidently of the opinion that while it might be a misfortune to be an old man, it was a fault to be a young one and good looking where girls were concerned.

'Miss Gale's father and mother showed me so much kindness,' Tom explained, seizing his moment of advantage, 'I should like to be of some use: it may not be convenient for you to come into town again in this cold weather.'

'Our folks have plenty to do all the time, that's a fact,' acknowledged uncle Ezra less grimly, while Nancy managed to show the light of a very knowing little smile. 'I don't know but she'd like to have a city man show her about, anyways. 'T ain't but four miles an' a half out to our place, the way we come, but while this weather holds I don't calculate to get into Boston more 'n once a week. I fetch all my stuff in to the Quincy Market myself, an' I've got to come in day after to-morrow mornin', but not till late, with a barrel o' nice winter pears I've been a-savin'. I can set the barrel right for'ard in the sleigh here, and I do' know but I can fetch Nancy as well as not. But how'd ye get home, Nancy? Could ye walk over to our place from the Milton depot, or couldn't ye?'

'Why, of course I could!' answered his niece, with a joy calmed by discretion.

' 'T ain't but a mile an' three quarters; 't won't hurt a State o' Maine girl,' said the old man, smiling under his great cap, so that his cold, shrewd eyes suddenly grew blue and boyish. 'I know all about ye now, Mr Aldis; I used to be well acquainted with your grandfather. Much obliged to you. Yes, I'll fetch Nancy. I'll leave her right up there to the Missionary Building, corner o' Somerset Street.* She can wait in the bookstore; it's liable to be open early. After I get through business to-day, I'm goin' to leave the hoss, an' let her see Faneuil Hall;* an' the market o' course, and I don't know but we shall stop in to the Old South Church;* or you can show her that, an' tell her about any other curiosities, if we don't have time.'

Nancy looked radiant, and Tom Aldis accepted his trust with satisfaction. At that moment the blockade was over and teams began to move.

'Not if it rains!' said uncle Ezra, speaking distinctly over his shoulder as they started. 'Otherwise expect her about eight or a little'——but the last of the sentence was lost.

Nancy looked back and nodded from the tangle to Tom, who stood on the curbstone with his hands in his pockets. Her white hood bobbed out of sight the next moment in School Street behind a great dray.

'Good gracious! eight o'clock!' said Tom, a little daunted, as he walked quickly up the street. As he passed the Missionary Building and the bookstore, he laughed aloud; but as he came near the clubhouse again, in this victorious retreat, he looked up at a window of one of the pleasant old houses, and then obeyed the beckoning nod of an elderly relative who seemed to have been watching for his return.

'Tom,' said she, as he entered the library, 'I insist upon it that I am not curious by nature or by habit, but what in the world made you chase that funny old horse and sleigh?'

'A pretty girl,' said Tom frankly.

II

THE second morning after this unexpected interview was sunshiny enough, and as cold as January could make it. Tom

Aldis, being young and gay, was apt to keep late hours at this season, and the night before had been the night of a Harvard assembly.* He was the kindest-hearted fellow in the world, but it was impossible not to feel a little glum and sleepy as he hurried toward the Missionary Building. The sharp air had urged uncle Ezra's white horse beyond his customary pace, so that the old sleigh was already waiting, and uncle Ezra himself was flapping his chilled arms and tramping to and fro impatiently.

'Cold mornin'!' he said. 'She's waitin' for you in there. I wanted to be sure you'd come. Now I'll be off. I've got them pears well covered, but I expect they may be touched. Nancy counted on comin', an' I'd just as soon she'd have a nice time. Her cousin's folks'll see her to the depot,' he added as he drove away, and Tom nodded reassuringly from the book-store door.

Nancy looked up eagerly from beside a counter full of gayly bound books, and gave him a speechless and grateful good-morning.

'I'm getting some presents for the little boys,' she informed him. 'They're great hands to read. This one's all about birds, for Sam, and I don't know but this Life o' Napoleon'll please Asa as much as anything. When I waked up this morning I felt homesick. I couldn't see anything out o' the window that I knew. I'm a real home body.'

'I should like to send the boys a present, myself,' said Tom. 'What do you think about jack-knives?'

'Asa'd rather have readin' matter; he ain't got the use for a knife that some boys have. Why, you're real good!' said Nancy.

'And your mother,—can't I send her something that she would like?' asked Tom kindly.

'She liked all those things that you and Mr Carew sent at Christmas time. We had the loveliest time opening the bundles. You oughtn't to think o' doing anything more. I wish you'd help me pick out a nice large-print Bible for grandma; she's always wishing for a large-print Bible, and her eyes fail her a good deal.'

Tom Aldis was not very fond of shopping, but this pious errand did not displease him in Nancy's company. A few minutes later, when they went out into the cold street, he felt warm and cheerful, and carried under his arm the flat parcel which held a large-print copy of the Scriptures and the little boys' books. Seeing Nancy again seemed to carry his thoughts back to East Rodney, as if he had been born and brought up there as well as she. The society and scenery of the little coast town were so simple and definite in their elements that one easily acquired a feeling of citizenship; it was like becoming acquainted with a friendly individual. Tom had an intimate knowledge, gained from several weeks' residence, with Nancy's whole world.

The long morning stretched before them like a morning in far Cathay, and they stepped off down the street toward the Old South Church, which had been omitted from uncle Ezra's scheme of entertainment by reason of difficulty in leaving the horse. The discovery that the door would not be open for nearly another hour only involved a longer walk among the city streets, and the asking and answering of many questions about the East Rodney neighbors, and the late autumn hunting and fishing which, with some land interests of his father's, had first drawn Tom to that part of the country. He had known enough of the rest of the world to appreciate the little community of fishermen-farmers, and while his friend Carew was but a complaining captive with a sprained ankle, Tom Aldis entered into the spirit of rural life with great zest; in fact he now remembered some boyish gallantries with a little uneasiness, and looked to Nancy to befriend him. It was easy for a man of twenty-two to arrive at an almost brotherly affection for such a person as Nancy; she was so discreet and so sincerely affectionate.

Nancy looked up at him once or twice as they walked along, and her face glowed with happy pride. 'I'd just like to have Addie Porter see me now!' she exclaimed, and gave Tom a straightforward look to which he promptly responded.

'Why?' he asked.

Nancy drew a long breath of relief, and began to smile.

'Oh, nothing,' she answered; 'only she kept telling me that you wouldn't have much of anything to say to me, if I should happen to meet you anywhere up to Boston. I knew better. I guess you're all right, aren't you, about that?' She spoke with sudden impulse, but there was something in her tone that made Tom blush a little.

'Why, yes,' he answered. 'What do you mean, Nancy?'

'We won't talk about it now while we're full of seeing things, but I've got something to say by and by,' said the girl soberly.

'You're very mysterious,' protested Tom, taking the bundle under his other arm, and piloting her carefully across the street.

Nancy said no more. The town was more interesting now that it seemed to have waked up, and her eyes were too busy. Everything proved delightful that day, from the recognition of business signs familiar to her through newspaper advertisements, to the Great Organ,* and the thrill which her patriotic heart experienced in a second visit to Faneuil Hall. They found the weather so mild that they pushed on to Charlestown, and went to the top of the monument,* which Tom had not done since he was a very small boy. After this they saw what else they could of historic Boston, on the fleetest and lightest of feet, and talked all the way, until they were suddenly astonished to hear the bells in all the steeples ring at noon.

'Oh dear, my nice mornin' 's all gone,' said Nancy regretfully. 'I never had such a beautiful time in all my life!'

She looked quite beautiful herself as she spoke: her eyes shone with lovely light and feeling, and her cheeks were bright with color like a fresh-bloomed rose, but for the first time that day she was wistful and sorry.

'Oh, you needn't go back yet!' said Tom. 'I've nothing in the world to do.'

'Uncle Ezra thought I'd better go up to cousin Snow's in Revere Street. I'm afraid she'll be all through dinner, but never mind. They thought I'd better go there on mother's account; it's her cousin, but I never saw her, at least not since I can remember. They won't like it if I don't, you know; it wouldn't be very polite.'

'All right,' assented Tom with dignity. 'I'll take you there at once: perhaps we can catch a car or something.'

'I'm ashamed to ask for anything more when you've been so kind,' said Nancy, after a few moments of anxious silence. 'I don't know that you can think of any good chance, but I'd give a great deal if I could only go somewhere and see some pretty dancing. You know I'm always dreamin' and dreamin' about pretty dancing!' and she looked eagerly at Tom to see what he would say. 'It must be goin' on somewhere in Boston,' she went on with pleading eyes. 'Could you ask somebody? They said at uncle Ezra's that if cousin Abby Snow wanted me to remain until to-morrow it might be just as well to stay; she used to be so well acquainted with mother. And so I thought—I might get some nice chance to look on.'

'To see some dancing,' repeated Tom, mindful of his own gay evening the night before, and of others to come, and the general impossibility of Nancy's finding the happiness she sought. He never had been so confronted by social barriers. As for Nancy's dancing at East Rodney, in the schoolhouse hall or in Jacob Parker's new barn, it had been one of the most ideal things he had ever known in his life; it would be hard to find elsewhere such grace as hers. In seaboard towns one often comes upon strange foreign inheritances, and the soul of a Spanish grandmother might still survive in Nancy, as far as her light feet were concerned. She danced like a flower in the wind. She made you feel light of foot yourself, as if you were whirling and blowing and waving through the air; as if you could go out dancing and dancing over the deep blue sea water of the bay, and find floor enough to touch and whirl upon. But Nancy had always seemed to take her gifts for granted; she had the simplicity of genius. 'I can't say now, but I am sure to find out,' said Tom Aldis definitely. 'I'll try to make some sort of plan for you. I wish we could have another dance, ourselves.'

'Oh, not now,' answered Nancy sensibly. 'It's knowing 'most all the people that makes a party pleasant.'

'My aunt would have asked you to come to luncheon to-day, but she had to go out of town, and was afraid of not getting back in season. She would like to see you very much. You see, I'm only a bachelor in lodgings, this winter,' explained Tom bravely.

'You've been just as good as you could be. I know all about Boston now, almost as if I lived here. I should like to see the inside of one of those big houses,' she added softly; 'they all look so noble as you go by. I think it was very polite of your aunt; you must thank her, Mr Aldis.'

It seemed to Tom as if his companion were building most glorious pleasure out of very commonplace materials. All the morning she had been as gay and busy as a brook.

By the middle of the afternoon he knocked again at cousin Snow's door in Revere Street, and delivered an invitation. Mrs Annesley, his aunt, and the kindest of women, would take Nancy to an afternoon class at Papanti's,* and bring her back afterwards, if cousin Snow were willing to spare her. Tom would wait and drive back with her in the coupé; then he must hurry to Cambridge for a business meeting to which he had been suddenly summoned.

Nancy was radiant when she first appeared, but a few minutes later, as they drove away together, she began to look grave and absent. It was only because she was so sorry to think of parting.

'I am so glad about the dancing class,' said Tom. 'I never should have thought of that. They are all children, you know; but it's very pretty, and they have all the new dances. I used to think it a horrid penance when I was a small boy.'

'I don't know why it is,' said Nancy, 'but the mere thought of music and dancin' makes me feel happy. I never saw any real good dancin', either, but I can always think what it ought to be. There's nothing so beautiful to me as manners,' she added softly, as if she whispered at the shrine of confidence.

'My aunt thinks there are going to be some pretty figure dances to-day,' announced Tom in a matter-of-fact way. There was something else than the dancing upon his mind. He thought that he ought to tell Nancy of his engagement,— not that it was quite an engagement yet,—but he could not do it just now. 'What was it you were going to tell me this morning? About Addie Porter, wasn't it?' He laughed a little, and then colored deeply. He had been somewhat foolish in his attentions to this young person, the beguiling village belle of East Rodney and the adjacent coasts. She was a pretty creature and a sad flirt, with none of the real beauty and quaint sisterly ways of Nancy. 'What was it all about?' he asked again.

Nancy turned away quickly. 'That's one thing I wanted to come to Boston for; that's what I want to tell you. She don't really care anything about you. She only wanted to get you away from the other girls. I know for certain that she likes Joe Brown better than anybody, and now she's been going with him almost all winter long. He keeps telling round that they're going to be married in the spring; but I thought if they were, she'd ask me to get some of her best things while I was in Boston. I suppose she's intendin' to play with him a while longer,' said Nancy with honest scorn, 'just because he loves her well enough to wait. But don't you worry about her, Mr Aldis!'

'I won't indeed,' answered Tom meekly, but with an unexpected feeling of relief as if the unconscious danger had been a real one. Nancy was very serious.

'I'm going home the first of the week,' she said as they parted; but the small hand felt colder than usual, and did not return his warm grasp. The light in her eyes had all gone, but Tom's beamed affectionately.

'I never thought of Addie Porter afterward, I'm afraid,' he confessed. 'What awfully good fun we all had! I should like to go down to East Rodney again some time.'

'Oh, shan't you ever come?' cried Nancy, with a thrill in her voice which Tom did not soon forget. He did not know that the young girl's heart was waked, he was so busy with the affairs of his own affections; but true friendship does not grow on every bush, in Boston or East Rodney, and Nancy's voice and farewell look touched something that lay very deep within his heart.

There is a little more to be told of this part of the story. Mrs Annesley, Tom's aunt, being a woman whose knowledge of human nature and power of sympathy made her a woman of the world rather than of any smaller circle,—Mrs Annesley was delighted with Nancy's unaffected pleasure and self-forgetful dignity of behavior at the dancing-school. She took her back to the fine house, and they had half an hour together there, and only parted because Nancy was to spend the night with cousin Snow, and another old friend of her mother's was to be asked to tea. Mrs Annesley asked her to come to see her again, whenever she was in Boston, and Nancy gratefully promised, but she never came. 'I'm all through with Boston for this time,' she said, with an amused smile, at parting. 'I'm what one of our neighbors calls "all flustered up,"' and she looked eagerly in her new friend's kind eyes for sympathy. 'Now that I've seen this beautiful house, and you and Mr Aldis, and some pretty dancin', I want to go right home where I belong.'

Tom Aldis meant to write to Nancy when his engagement came out, but he never did; and he meant to send a long letter to her and her mother two years later, when he and his wife were going abroad for a long time; but he had an inborn hatred of letter-writing, and let that occasion pass also, though when anything made him very sorry or very glad, he had a curious habit of thinking of these East Rodney friends. Before he went to Europe he used to send them magazines now and then, or a roll of illustrated papers; and one day, in a bookstore, he happened to see a fine French book with colored portraits of famous dancers, and sent it by express to Nancy with his best remembrances. But Tom was young and much occupied, the stream of time floated him away from the shore of Maine, not toward it, ten or fifteen years passed by, his brown hair began to grow gray, and he came back from Europe after a while to a new Boston life in which reminiscences of East Rodney seemed very remote indeed.

III

ONE summer afternoon there were two passengers, middle-aged men, on the small steamer James Madison, which attended the comings and goings of the great Boston steamer, and ran hither and yon on errands about Penobscot Bay. She was puffing up a long inlet toward East Rodney Landing, and the two strangers were observing the green shores with great interest. Like nearly the whole stretch of the Maine coast, there was a house on almost every point and headland; but for all this, there were great tracts of untenanted country, dark untouched forests of spruces and firs, and shady coves where there seemed to be deep water and proper moorings. The two passengers were on the watch for landings and lookouts; in short, this lovely, lonely country was being frankly appraised at its probable value for lumbering or for building-lots and its relation to the real estate market. Just now there appeared to be no citizens save crows and herons, the sun was almost down behind some high hills in the west, and the Landing was in sight not very far ahead.

'It is nearly twenty years since I came down here before,' said the younger of the two men, suddenly giving the conversation a personal turn. 'Just after I was out of college, at any rate. My father had bought this point of land with the islands. I think he meant to come and hunt in the autumn, and was misled by false accounts of deer and moose. He sent me down to oversee something or other; I believe he had some surveyors at work, and thought they had better be looked after; so I got my chum Carew to come along, and we found plenty of trout, and had a great time until he gave his ankle a bad sprain.'

'What did you do then?' asked the elder man politely, keeping his eyes on the shore.

'I stayed by, of course; I had nothing to do in those days,' answered Mr Aldis. 'It was one of those nice old-fashioned country neighborhoods where there was plenty of fun among the younger people,—sailing on moonlight nights, and hay-cart parties, and dances, and all sorts of things. We used to go to prayer-meeting nine or ten miles off, and sewing societies. I had hard work to get away! We made excuse of Carew's ankle joint as long as we could, but he'd been all right and going everywhere with the rest of us a fortnight before we started. We waited until there was ice alongshore, I remember.'

'Daniel R. Carew, was it, of the New York Stock Exchange?' asked the listener. 'He strikes you as being a very grave sort of person now; doesn't like it if he finds anybody in his chair at the club, and all that.'

'I can stir him up,' said Mr Aldis confidently. 'Poor old fellow, he has had a good deal of trouble, one way and another. How the Landing has grown up! Why, it's a good-sized little town!'

'I'm sorry it is so late,' he added, after a long look at a farm on the shore which they were passing. 'I meant to go to see the people up there,' and he pointed to the old farmhouse, dark and low and firm-rooted in the long slope of half-tamed, ledgy fields. Warm thoughts of Nancy filled his heart, as if they had said good-by to each other that cold afternoon in Boston only the winter before. He had not been so eager to see any one for a long time. Such is the triumph of friendship: even love itself without friendship is the victim of chance and time.

When supper was over in the Knox House, the one centre of public entertainment in East Rodney, it was past eight o'clock, and Mr Aldis felt like a dim copy of Rip Van Winkle, or of the gay Tom Aldis who used to know everybody, and be known of all men as the planner of gayeties. He lighted a cigar as he sat on the front piazza of the hotel, and gave himself up to reflection. There was a long line of lights in the second story of a wooden building opposite, and he was conscious of some sort of public interest and excitement.

'There is going to be a time in the hall,' said the landlord, who came hospitably out to join him. 'The folks are going to have a dance. The proceeds will be applied to buying a bell for the new schoolhouse. They'd be pleased if you felt like stepping over; there has been a considerable number glad to hear you thought of coming down. I ain't an East Rodney man myself, but I've often heard of your residin' here some years ago. Our folks is makin' the ice cream for the occasion,' he added significantly, and Mr Aldis nodded and smiled in acknowledgment. He had meant to go out and see the Gales, if the boat had only got in in season; but boats are unpunctual in their ways, and the James Madison had been unexpectedly signaled by one little landing and settlement after another. He remembered that a great many young people were on board when they arrived, and now they appeared again, coming along the street and disappearing at the steep stairway opposite. The lighted windows were full of heads already, and there were now and then preliminary exercises upon a violin.

Mr Aldis had grown old enough to be obliged to sit and think
it over about going to a ball; the day had passed when there
would have been no question; but when he had finished his
cigar he crossed the street, and only stopped before the lighted store window to find a proper bank bill for the door-
keeper. Then he ran up the stairs to the hall, as if he were the
Tom Aldis of old. It was an embarrassing moment as he
entered the low, hot room, and the young people stared at
him suspiciously; but there were also elderly people scattered
about who were meekly curious and interested, and one of
these got clumsily upon his feet and hastened to grasp the
handsome stranger by the hand.

'Nancy heard you was coming,' said Mr Gale delightedly.
'She expected I should see you here, if you was just the same
kind of a man you used to be. Come let's set right down, folks
is crowding in; there may be more to set than there is to
dance.'

'How is Nancy, isn't she coming?' asked Tom, feeling the
years tumble off his shoulders.

'Well as usual, poor creatur',' replied the old father, with a
look of surprise. 'No, no; she can't go nowhere.'

At that moment the orchestra struck up a military march
with so much energy that further conversation was impossible. Near them was an awkward-looking young fellow, with
shoulders too broad for his height, and a general look of
chunkiness and dullness. Presently he rose and crossed the
room, and made a bow to his chosen partner that most
courtiers might have envied. It was a bow of grace and dignity.

'Pretty well done!' said Tom Aldis aloud.

Mr Gale was beaming with smiles, and keeping time to the
music with his foot and hand. 'Nancy done it,' he announced
proudly, speaking close to his companion's ear. 'That boy
give her a sight o' difficulty; he used to want to learn, but
'long at the first he'd turn red as fire if he much as met a
sheep in a pastur'. The last time I see him on the floor I went
home an' told her he done as well as any. You can see for
yourself, now they're all a-movin'.'

The fresh southerly breeze came wafting into the hall and
making the lamps flare. If Tom turned his head, he could see

the lights out in the bay, of vessels that had put in for the night. Old Mr Gale was not disposed for conversation so long as the march lasted, and when it was over a frisky-looking middle-aged person accosted Mr Aldis with the undimmed friendliness of their youth; and he took her out, as behoved him, for the Lancers quadrille.* From her he learned that Nancy had been for many years a helpless invalid; and when their dance was over he returned to sit out the next one with Mr Gale, who had recovered a little by this time from the excitement of the occasion, and was eager to talk about Nancy's troubles, but still more about her gifts and activities. After a while they adjourned to the hotel piazza in company, and the old man grew still more eloquent over a cigar. He had not changed much since Tom's residence in the family; in fact, the flight of seventeen years had made but little difference in his durable complexion or the tough frame which had been early seasoned by wind and weather.

'Yes, sir,' he said, 'Nancy has had it very hard, but she's the life o' the neighborhood yet. For excellent judgment I never see her equal. Why, once the board o' selec'men took trouble to meet right there in her room off the kitchen, when they had to make some responsible changes in layin' out the school deestricts. She was the best teacher they ever had, a master good teacher; fitted a boy for Bowdoin College all except his Greek, that last season before she was laid aside from sickness. She took right holt to bear it the best she could, and begun to study on what kind o' things she could do. First she used to make out to knit, a-layin' there, for the store, but her hands got crippled up with the rest of her; 't is the wust kind o' rheumatics there is. She had me go round to the neighborin' schools and say that if any of the child'n was backward an' slow with their lessons to send 'em up to her. Now an' then there'd be one, an' at last she'd see to some class there wasn't time for: an' here year before last the town voted her fifty dollars a year for her services. What do you think of that?'

Aldis manifested his admiration, but he could not help wishing that he had not seemed to forget so pleasant an old acquaintance, and above all wished that he had not seemed to take part in nature's great scheme to defraud her. She had begun life with such distinct rights and possibilities.

'I tell you she was the most cut up to have to stop dancin',' said Mr Gale gayly, 'but she held right on to that, same as to other things. "I can't dance myself," she says, "so I'm goin' to make other folks." You see right before you how she's kep' her word, Mr Aldis? What always pleased her the most, from a child, was dancin'. Folks talked to her some about letting her mind rove on them light things when she appeared to be on a dyin' bed. "David, he danced afore the Lord,"* she'd tell 'em, an' her eyes would snap so, they didn't like to say no more.'

Aldis laughed, the old man himself was so cheerful.

'Well, sir, she made 'em keep right on with the old dancin'-school she always took such part in (I guess 't was goin', wa'n't it, that fall you stopped here?); but she sent out for all the child'n she could get and learnt 'em their manners. She can see right out into the kitchen from where she is, an' she has 'em make their bows an' take their steps till they get 'em right an' feel as good as anybody. There's boys an' girls comin' an' goin' two or three times a week in the afternoon. It don't seem to be no hardship: there ain't no such good company for young or old as Nancy.'

'She'll be dreadful glad to see you,' the proud father ended his praises. 'Oh, she's never forgot that good time she had up to Boston. You an' all your folks couldn't have treated her no better, an' you give her her heart's desire, you did so! She's never done talkin' about that pretty dancin'-school with all them lovely little child'n, an' everybody so elegant and pretty behaved. She'd always wanted to see such a lady as your aunt was. I don't know but she's right: she always maintains that when folks has good manners an' good hearts the world is their 'n, an' she was goin' to do everything she could to keep young folks from feelin' hoggish an' left out.'

Tom walked out toward the farm in the bright moonlight with Mr Gale, and promised to call as early the next day as possible. They followed the old shore path, with the sea on one side and the pointed firs on the other, and parted where Nancy's light could be seen twinkling on the hill.

IV

IT was not very cheerful to look forward to seeing a friend of one's youth crippled and disabled; beside, Tom Aldis always felt a nervous dread in being where people were ill and suffering. He thought once or twice how little compassion for Nancy these country neighbors expressed. Even her father seemed inclined to boast of her, rather than to pity the poor life that was so hindered. Business affairs and conference were appointed for that afternoon, so that by the middle of the morning he found himself walking up the yard to the Gales' side door.

There was nobody within call. Mr Aldis tapped once or twice, and then hearing a voice he went through the narrow unpainted entry into the old kitchen, a brown, comfortable place which he well remembered.

'Oh, I'm so glad to see you,' Nancy was calling from her little bedroom beyond. 'Come in, come in!'

He passed the doorway, and stood with his hand on hers, which lay helpless on the blue-and-white coverlet. Nancy's young eyes, untouched by years or pain or regret, looked up at him as frankly as a child's from the pillow.

'Mother's gone down into the field to pick some peas for dinner,' she said, looking and looking at Tom and smiling; but he saw at last that tears were shining, too, and making her smile all the brighter. 'You see now why I couldn't write,' she explained. 'I kept thinking I should. I didn't want anybody else to thank you for the books. Now sit right down,' she begged her guest. 'Father told me all he could about last night. You danced with Addie Porter.'

'I did,' acknowledged Tom Aldis, and they both laughed. 'We talked about old times between the figures, but it seemed to me that I remembered them better than she did.'

'Addie has been through with a good deal of experience since then,' explained Nancy, with a twinkle in her eyes.

'I wish I could have danced again with you,' said Tom bravely, 'but I saw some scholars that did you credit.'

'I have to dance by proxy,' said Nancy; and to this there was no reply.

Tom Aldis sat in the tiny bedroom with an aching heart.

Such activity and definiteness of mind, such power of loving and hunger for life, had been pent and prisoned there so many years. Nancy had made what she could of her small world of books. There was something very uncommon in her look and way of speaking; he felt like a boy beside her,—he to whom the world had given its best luxury and widest opportunity. As he looked out of the small window, he saw only a ledgy pasture where sheep were straying along the slopes among the bayberry and juniper; beyond were some balsam firs and a glimpse of the sea. It was a lovely bit of landscape, but it lacked figures, and Nancy was born to be a teacher and a lover of her kind. She had only lacked opportunity, but she was equal to meeting whatever should come. One saw it in her face.

'You don't know how many times I have thought of that cold day in Boston,' said Nancy from her pillows. 'Your aunt was beautiful. I never could tell you about the rest of the day with her, could I? Why, it just gave me a measure to live by. I saw right off how small some things were that I thought were big. I told her about one or two things down here in Rodney that troubled me, and she understood all about it. "If we mean to be happy and useful," she said, "the only way is to be self-forgetful." I never forgot that!'

'The seed fell upon good ground,* didn't it?' said Mr Aldis with a smile. He had been happy enough himself, but Nancy's happiness appeared in that moment to have been of another sort. He could not help thinking what a wonderful perennial quality there is in friendship. Because it had once flourished and bloomed, no winter snows of Maine could bury it, no summer sunshine of foreign life could wither this single flower of a day long past. The years vanished like a May snowdrift, and because they had known each other once they found each other now. It was like a tough little sprig of gray everlasting; the New England edelweiss that always keeps a white flower ready to blossom safe and warm in its heart.

They entertained each other delightfully that late summer morning. Tom talked of his wife and children as he had seldom talked of them to any one before, and afterward explained the land interests which had brought him back at this late day to East Rodney.

'I came down meaning to sell my land to a speculator,' he said, 'or to a real estate agency which has great possessions along the coast; but I'm very doubtful about doing it, now that I have seen the bay again and this lovely shore. I had no idea that it was such a magnificent piece of country. I was going on from here to Mount Desert, with a half idea of buying land there. Why isn't this good enough that I own already? With a yacht or a good steam launch we shouldn't be so far away from places along the coast, you know. What if I were to build a house above Sunday Cove, on the headland, and if we should be neighbors! I have a friend who might build another house on the point beyond; we came home from abroad at about the same time, and he's looking for a place to build, this side of Bar Harbor.' Tom was half confiding in his old acquaintance, and half thinking aloud. 'These real estate brokers can't begin to give a man the value of such land as mine,' he added.

'It would be excellent business to come and live here yourself, if you want to bring up the value of the property,' said Nancy gravely. 'I hear there are a good many lots staked out between here and Portland, but it takes more than that to start things. There can't be any prettier place than East Rodney,' she declared, looking affectionately out of her little north window. 'It would be a great blessing to city people, if they could come and have our good Rodney air.'

The friends talked on a little longer, and with great cheerfulness and wealth of reminiscence. Tom began to understand why nobody seemed to pity Nancy, though she did at last speak sadly, and make confession that she felt it to be very hard because she never could get about the neighborhood to see any of the old and sick people. Some of them were lonesome, and lived in lonesome places. 'I try to send word to them sometimes, if I can't do any more,' said Nancy. 'We're so apt to forget 'em, and let 'em feel they aren't useful. I can't bear to see an old heart begging for a little love. I do sometimes wish I could manage to go an' try to make a little of their time pass pleasant.'

'Do you always stay just here?' asked Tom with sudden compassion, after he had stood for a moment looking out at the gray sheep on the hillside.

'Oh, sometimes I get into the old rocking-chair, and father pulls me out into the kitchen when I'm extra well,' said Nancy proudly, as if she spoke of a yachting voyage or a mountaineer's exploits. 'Once a doctor said if I was only up to Boston'—her voice fell a little with a touch of wistfulness— 'perhaps I could have had more done, and could have got about with some kind of a chair. But that was a good while ago: I never let myself worry about it. I am so busy right here that I don't know what would happen if I set out to travel.'

V

A YEAR later the East Rodney shore looked as green as ever, and the untouched wall of firs and pines faithfully echoed the steamer's whistle. In the twelve months just past Mr Aldis had worked wonders upon his long-neglected estate, and now was comfortably at housekeeping on the Sunday Cove headland. Nancy could see the chimneys and a gable of the fine establishment from her own little north window, and the sheep still fed undisturbed on the slopes that lay between. More than this, there were two other new houses, to be occupied by Tom's friends, within the distance of a mile or two. It would be difficult to give any idea of the excitement and interest of East Rodney, or the fine effect and impulse to the local market. Tom's wife and children were most affectionately befriended by their neighbors the Gales, and with their coming in midsummer many changes for the better took place in Nancy's life, and made it bright. She lost no time in starting a class, where the two eldest for the first time found study a pleasure, while little Tom was promptly and tenderly taught his best bow, and made to mind his steps with such interest and satisfaction that he who had once roared aloud in public at the infant dancing-class, now knew both confidence and ambition. There was already a well-worn little footpath between the old Gale house and Sunday Cove; it wound in and out among the ledges and thickets, and over the short sheep-turf of the knolls; and there was a scent of sweet-brier here, and of raspberries there, and of the salt water and the pines, and the juniper and bayberry, all the way.

Nancy herself had followed that path in a carrying-chair, and joy was in her heart at every step. She blessed Tom over and over again, as he walked, broad-shouldered and strong, between the forward handles, and turned his head now and then to see if she liked the journey. For many reasons, she was much better now that she could get out into the sun. The bedroom with the north window was apt to be tenantless, and wherever Nancy went she made other people wiser and happier, and more interested in life.

On the day when she went in state to visit the new house, with her two sober carriers, and a gay little retinue of young people frisking alongside, she felt happy enough by the way; but when she got to the house itself, and had been carried quite round it, and was at last set down in the wide hall to look about, she gave her eyes a splendid liberty of enjoyment. Mrs Aldis disappeared for a moment to give directions in her guest's behalf, and the host and Nancy were left alone together.

'No, I don't feel a bit tired,' said the guest, looking pale and radiant. 'I feel as if I didn't know how to be grateful enough. I have everything in the world to make me happy. What does make you and your dear family do so much?'

'It means a great deal to have friends, doesn't it?' answered Tom in a tone that thanked her warmly. 'I often wish'——

He could not finish his sentence, for he was thinking of Nancy's long years, and the bond of friendship that absence and even forgetfulness had failed to break; of the curious insistence of fate which made him responsible for something in the life of Nancy and brought him back to her neighborhood. It was a moment of deep thought; he even forgot Nancy herself. He heard the water plashing on the shore below, and felt the cool sea wind that blew in at the door.

Nancy reached out her bent and twisted hand and began to speak; then she hesitated, and glanced at her hand again, and looked straight at him with shining eyes.

'There never has been a day when I haven't thought of you,' she said.

MARTHA'S LADY

I

ONE day, many years ago, the old Judge Pyne house wore an unwonted look of gayety and youthfulness. The high-fenced green garden was bright with June flowers. Under the elms in the large shady front yard you might see some chairs placed near together, as they often used to be when the family were all at home and life was going on gayly with eager talk and pleasure-making; when the elder judge, the grandfather, used to quote that great author, Dr Johnson,* and say to his girls, 'Be brisk, be splendid, and be public.'

One of the chairs had a crimson silk shawl thrown carelessly over its straight back, and a passer-by, who looked in through the latticed gate between the tall gate-posts with their white urns, might think that this piece of shining East Indian color was a huge red lily that had suddenly bloomed against the syringa bush. There were certain windows thrown wide open that were usually shut, and their curtains were blowing free in the light wind of a summer afternoon; it looked as if a large household had returned to the old house to fill the prim best rooms and find them full of cheer.

It was evident to every one in town that Miss Harriet Pyne, to use the village phrase, had company. She was the last of her family, and was by no means old; but being the last, and wonted to live with people much older than herself, she had formed all the habits of a serious elderly person. Ladies of her age, something past thirty, often wore discreet caps in those days, especially if they were married, but being single, Miss Harriet clung to youth in this respect, making the one concession of keeping her waving chestnut hair as smooth and stiffly arranged as possible. She had been the dutiful companion of her father and mother in their latest years, all her elder brothers and sisters having married and gone, or died and gone, out of the old house. Now that she was left alone it seemed quite the best thing frankly to accept the fact of age, and to turn more resolutely than ever to the companionship of duty and serious books. She was more serious and given to routine than her elders themselves, as sometimes happened when the daughters of New England gentlefolks were brought up wholly in the society of their elders. At thirty-five she had more reluctance than her mother to face an unforeseen occasion, certainly more than her grandmother, who had preserved some cheerful inheritance of gayety and worldliness from colonial times.

There was something about the look of the crimson silk shawl in the front yard to make one suspect that the sober customs of the best house in a quiet New England village were all being set at defiance, and once when the mistress of the house came to stand in her own doorway, she wore the pleased but somewhat apprehensive look of a guest. In these days New England life held the necessity of much dignity and discretion of behavior; there was the truest hospitality and good cheer in all occasional festivities, but it was sometimes a self-conscious hospitality, followed by an inexorable return to asceticism both of diet and of behavior. Miss Harriet Pyne belonged to the very dullest days of New England, those which perhaps held the most priggishness for the learned professions, the most limited interpretation of the word 'evangelical,' and the pettiest indifference to large things. The outbreak of a desire for larger religious freedom caused at first a most determined reaction toward formalism, especially in small and quiet villages like Ashford, intently busy with their own concerns. It was high time for a little leaven to begin its work, in this moment when the great impulses of the war for liberty had died away and those of the coming war for patriotism and a new freedom had hardly yet begun.

The dull interior, the changed life of the old house, whose former activities seemed to have fallen sound asleep, really typified these larger conditions, and a little leaven had made its easily recognized appearance in the shape of a light-hearted girl. She was Miss Harriet's young Boston cousin, Helena Vernon, who, half-amused and half-impatient at the unnecessary sober-mindedness of her hostess and of Ashford in general, had set herself to the difficult task of gayety. Cousin Harriet looked on at a succession of ingenious and, on the whole, innocent attempts at pleasure, as she might have looked on at the frolics of a kitten who easily substitutes a ball of yarn for the uncertainties of a bird or a wind-blown leaf, and who may at any moment ravel the fringe of a sacred curtain-tassel in preference to either.

Helena, with her mischievous appealing eyes, with her enchanting old songs and her guitar, seemed the more delightful and even reasonable because she was so kind to everybody, and because she was a beauty. She had the gift of most charming manners. There was all the unconscious lovely ease and grace that had come with the good breeding of her city home, where many pleasant people came and went; she had no fear, one had almost said no respect, of the individual, and she did not need to think of herself. Cousin Harriet turned cold with apprehension when she saw the minister coming in at the front gate, and wondered in agony if Martha were properly attired to go to the door, and would by any chance hear the knocker; it was Helena who, delighted to have anything happen, ran to the door to welcome the Reverend Mr Crofton as if he were a congenial friend of her own age. She could behave with more or less propriety during the stately first visit, and even contrive to lighten it with modest mirth, and to extort the confession that the guest had a tenor voice, though sadly out of practice; but when the minister departed a little flattered, and hoping that he had not expressed himself too strongly for a pastor upon the poems of Emerson,* and feeling the unusual stir of gallantry in his proper heart, it was Helena who caught the honored hat of the late Judge Pyne from its last resting-place in the hall, and holding it securely in both hands, mimicked the minister's self-conscious entrance. She copied his pompous and anxious expression in the dim parlor in such delicious fashion that Miss Harriet, who could not always extinguish a ready spark of the original sin of humor, laughed aloud.

'My dear!' she exclaimed severely the next moment, 'I am ashamed of your being so disrespectful!' and then laughed again, and took the affecting old hat and carried it back to its place.

'I would not have had any one else see you for the world,' she said sorrowfully as she returned, feeling quite self-possessed again, to the parlor doorway; but Helena still sat in the minister's chair, with her small feet placed as his stiff boots had been, and a copy of his solemn expression before they came to speaking of Emerson and of the guitar. 'I wish I had asked him if he would be so kind as to climb the cherry-tree,' said Helena, unbending a little at the discovery that her cousin would consent to laugh no more. 'There are all those ripe cherries on the top branches. I can climb as high as he, but I can't reach far enough from the last branch that will bear me. The minister is so long and thin'—

'I don't know what Mr Crofton would have thought of you; he is a very serious young man,' said cousin Harriet, still ashamed of her laughter. 'Martha will get the cherries for you, or one of the men. I should not like to have Mr Crofton think you were frivolous, a young lady of your opportunities'—but Helena had escaped through the hall and out at the garden door at the mention of Martha's name. Miss Harriet Pyne sighed anxiously, and then smiled, in spite of her deep convictions, as she shut the blinds and tried to make the house look solemn again.

The front door might be shut, but the garden door at the other end of the broad hall was wide open upon the large sunshiny garden, where the last of the red and white peonies and the golden lilies, and the first of the tall blue larkspurs lent their colors in generous fashion. The straight box borders were all in fresh and shining green of their new leaves, and there was a fragrance of the old garden's inmost life and soul blowing from the honeysuckle blossoms on a long trellis. It was now late in the afternoon, and the sun was low behind great apple-trees at the garden's end, which threw their shadows over the short turf of the bleaching-green. The cherry-trees stood at one side in full sunshine, and Miss Harriet, who presently came to the garden steps to watch like a hen at the water's edge, saw her cousin's pretty figure in its white dress of India muslin hurrying across the grass. She was accompanied by the tall, ungainly shape of Martha the new maid, who, dull and indifferent to every one else, showed a surprising willingness and allegiance to the young guest.

'Martha ought to be in the dining-room, already, slow as she is; it wants but half an hour of tea-time,' said Miss Harriet, as she turned and went into the shaded house. It was Martha's duty to wait at table, and there had been many trying scenes and defeated efforts toward her education. Martha was certainly very clumsy, and she seemed the clumsier because she had replaced her aunt, a most skillful person, who had but lately married a thriving farm and its prosperous owner. It must be confessed that Miss Harriet was a most bewildering instructor, and that her pupil's brain was easily confused and prone to blunders. The coming of Helena had been somewhat dreaded by reason of this incompetent service, but the guest took no notice of frowns or futile gestures at the first tea-table, except to establish friendly relations with Martha on her own account by a reassuring smile. They were about the same age, and next morning, before cousin Harriet came down, Helena showed by a word and a quick touch the right way to do something that had gone wrong and been impossible to understand the night before. A moment later the anxious mistress came in without suspicion, but Martha's eyes were as affectionate as a dog's, and there was a new look of hopefulness on her face; this dreaded guest was a friend after all, and not a foe come from proud Boston to confound her ignorance and patient efforts.

The two young creatures, mistress and maid, were hurrying across the bleaching-green.

'I can't reach the ripest cherries,' explained Helena politely, 'and I think that Miss Pyne ought to send some to the minister. He has just made us a call. Why, Martha, you haven't been crying again!'

'Yes 'm,' said Martha sadly. 'Miss Pyne always loves to send something to the minister,' she acknowledged with interest, as if she did not wish to be asked to explain these latest tears.

'We'll arrange some of the best cherries in a pretty dish. I'll show you how, and you shall carry them over to the parsonage after tea,' said Helena cheerfully, and Martha accepted the embassy with pleasure. Life was beginning to hold moments of something like delight in the last few days.

'You'll spoil your pretty dress, Miss Helena,' Martha gave

shy warning, and Miss Helena stood back and held up her skirts with unusual care while the country girl, in her heavy blue checked gingham, began to climb the cherry-tree like a boy.

Down came the scarlet fruit like bright rain into the green grass.

'Break some nice twigs with the cherries and leaves together; oh, you're a duck, Martha!' and Martha, flushed with delight, and looking far more like a thin and solemn blue heron, came rustling down to earth again, and gathered the spoils into her clean apron.

That night at tea, during her hand-maiden's temporary absence, Miss Harriet announced, as if by way of apology, that she thought Martha was beginning to understand something about her work. 'Her aunt was a treasure, she never had to be told anything twice; but Martha has been as clumsy as a calf,' said the precise mistress of the house. 'I have been afraid sometimes that I never could teach her anything. I was quite ashamed to have you come just now, and find me so unprepared to entertain a visitor.'

'Oh, Martha will learn fast enough because she cares so much,' said the visitor eagerly. 'I think she is a dear good girl. I do hope that she will never go away. I think she does things better every day, cousin Harriet,' added Helena pleadingly, with all her kind young heart. The china-closet door was open a little way, and Martha heard every word. From that moment, she not only knew what love was like, but she knew love's dear ambitions. To have come from a stony hill-farm and a bare small wooden house, was like a cave-dweller's coming to make a permanent home in an art museum, such had seemed the elaborateness and elegance of Miss Pyne's fashion of life; and Martha's simple brain was slow enough in its processes and recognitions. But with this sympathetic ally and defender, this exquisite Miss Helena who believed in her, all difficulties appeared to vanish.

Later that evening, no longer homesick or hopeless, Martha returned from her polite errand to the minister, and stood with a sort of triumph before the two ladies, who were sitting in the front doorway, as if they were waiting for visitors,

Helena still in her white muslin and red ribbons, and Miss Harriet in a thin black silk. Being happily self-forgetful in the greatness of the moment, Martha's manners were perfect, and she looked for once almost pretty and quite as young as she was.

'The minister came to the door himself, and returned his thanks. He said that cherries were always his favorite fruit, and he was much obliged to both Miss Pyne and Miss Vernon. He kept me waiting a few minutes, while he got this book ready to send to you, Miss Helena.'

'What are you saying, Martha? I have sent him nothing!' exclaimed Miss Pyne, much astonished. 'What does she mean, Helena?'

'Only a few cherries,' explained Helena. 'I thought Mr Crofton would like them after his afternoon of parish calls. Martha and I arranged them before tea, and I sent them with our compliments.'

'Oh, I am very glad you did,' said Miss Harriet, wondering, but much relieved. 'I was afraid'—

'No, it was none of my mischief,' answered Helena daringly. 'I did not think that Martha would be ready to go so soon. I should have shown you how pretty they looked among their green leaves. We put them in one of your best white dishes with the openwork edge. Martha shall show you to-morrow; mamma always likes to have them so.' Helena's fingers were busy with the hard knot of a parcel.

'See this, cousin Harriet!' she announced proudly, as Martha disappeared round the corner of the house, beaming with the pleasures of adventure and success. 'Look! the minister has sent me a book: Sermons on *what*? Sermons—it is so dark that I can't quite see.'

'It must be his "Sermons on the Seriousness of Life;" they are the only ones he has printed, I believe,' said Miss Harriet, with much pleasure. 'They are considered very fine discourses. He pays you a great compliment, my dear. I feared that he noticed your girlish levity.'

'I behaved beautifully while he stayed,' insisted Helena. 'Ministers are only men,' but she blushed with pleasure. It was certainly something to receive a book from its author, and such a tribute made her of more value to the whole reverent household. The minister was not only a man, but a bachelor, and Helena was at the age that best loves conquest; it was at any rate comfortable to be reinstated in cousin Harriet's good graces.

'Do ask the kind gentleman to tea! He needs a little cheering up,' begged the siren in India muslin, as she laid the shiny black volume of sermons on the stone doorstep with an air of approval, but as if they had quite finished their mission.

'Perhaps I shall, if Martha improves as much as she has within the last day or two,' Miss Harriet promised hopefully. 'It is something I always dread a little when I am all alone, but I think Mr Crofton likes to come. He converses so elegantly.'

II

THESE were the days of long visits, before affectionate friends thought it quite worth while to take a hundred miles' journey merely to dine or to pass a night in one another's houses. Helena lingered through the pleasant weeks of early summer, and departed unwillingly at last to join her family at the White Hills,* where they had gone, like other households of high social station, to pass the month of August out of town. The happy-hearted young guest left many lamenting friends behind her, and promised each that she would come back again next year. She left the minister a rejected lover, as well as the preceptor of the academy, but with their pride unwounded, and it may have been with wider outlooks upon the world and a less narrow sympathy both for their own work in life and for their neighbors' work and hindrances. Even Miss Harriet Pyne herself had lost some of the unnecessary provincialism and prejudice which had begun to harden a naturally good and open mind and affectionate heart. She was conscious of feeling younger and more free, and not so lonely. Nobody had ever been so gay, so fascinating, or so kind as Helena, so full of social resource, so simple and undemanding in her friendliness. The light of her young life cast no shadow on either young or old companions, her pretty clothes never seemed to make other girls look dull or out of fashion. When she went away up the street in Miss Harriet's carriage to take the slow train toward Boston and the gayeties of the new Profile House,* where her mother waited impatiently with a group of Southern friends, it seemed as if there would never be any more picnics or parties in Ashford, and as if society had nothing left to do but to grow old and get ready for winter.

Martha came into Miss Helena's bedroom that last morning, and it was easy to see that she had been crying; she looked just as she did in that first sad week of homesickness and despair. All for love's sake she had been learning to do many things, and to do them exactly right; her eyes had grown quick to see the smallest chance for personal service. Nobody could be more humble and devoted; she looked years older than Helena, and wore already a touching air of care-taking.

'You spoil me, you dear Martha!' said Helena from the bed. 'I don't know what they will say at home, I am so spoiled.'

Martha went on opening the blinds to let in the brightness of the summer morning, but she did not speak.

'You are getting on splendidly, aren't you?' continued the little mistress. 'You have tried so hard that you make me ashamed of myself. At first you crammed all the flowers together, and now you make them look beautiful. Last night cousin Harriet was so pleased when the table was so charming, and I told her that you did everything yourself, every bit. Won't you keep the flowers fresh and pretty in the house until I come back? It's so much pleasanter for Miss Pyne, and you'll feed my little sparrows, won't you? They're growing so tame.'

'Oh, yes, Miss Helena!' and Martha looked almost angry for a moment, then she burst into tears and covered her face with her apron. 'I couldn't understand a single thing when I first came. I never had been anywhere to see anything, and Miss Pyne frightened me when she talked. It was you made me think I could ever learn. I wanted to keep the place, 'count of mother and the little boys; we're dreadful hard pushed. Hepsy has been good in the kitchen; she said she ought to have patience with me, for she was awkward herself when she first came.'

Helena laughed; she looked so pretty under the tasseled white curtains.

'I dare say Hepsy tells the truth,' she said. 'I wish you had told me about your mother. When I come again, some day we'll drive up country, as you call it, to see her. Martha! I wish you would think of me sometimes after I go away. Won't you promise?' and the bright young face suddenly grew grave. 'I have hard times myself; I don't always learn things that I ought to learn, I don't always put things straight. I wish you wouldn't forget me ever, and would just believe in me. I think it does help more than anything.'

'I won't forget,' said Martha slowly. 'I shall think of you every day.' She spoke almost with indifference, as if she had been asked to dust a room, but she turned aside quickly and pulled the little mat under the hot water jug quite out of its former straightness; then she hastened away down the long white entry, weeping as she went.

III

To lose out of sight the friend whom one has loved and lived to please is to lose joy out of life. But if love is true, there comes presently a higher joy of pleasing the ideal, that is to say, the perfect friend. The same old happiness is lifted to a higher level. As for Martha, the girl who stayed behind in Ashford, nobody's life could seem duller to those who could not understand; she was slow of step, and her eyes were almost always downcast as if intent upon incessant toil; but they startled you when she looked up, with their shining light. She was capable of the happiness of holding fast to a great sentiment, the ineffable satisfaction of trying to please one whom she truly loved. She never thought of trying to make other people pleased with herself; all she lived for was to do the best she could for others, and to conform to an ideal, which grew at last to be like a saint's vision, a heavely figure painted upon the sky.

On Sunday afternoons in summer, Martha sat by the window of her chamber, a low-storied little room, which looked into the side yard and the great branches of an elm-tree. She never sat in the old wooden rocking-chair except on Sundays like this; it belonged to the day of rest and to happy meditation. She wore her plain black dress and a clean white apron, and held in her lap a little wooden box, with a brass ring on top for a handle. She was past sixty years of age and looked even older, but there was the same look on her face that it had sometimes worn in girlhood. She was the same Martha; her hands were old-looking and work-worn, but her face still shone. It seemed like yesterday that Helena Vernon had gone away, and it was more than forty years.

War and peace had brought their changes and great anxieties, the face of the earth was furrowed by floods and fire, the faces of mistress and maid were furrowed by smiles and tears, and in the sky the stars shone on as if nothing had happened. The village of Ashford added a few pages to its unexciting history, the minister preached, the people listened; now and then a funeral crept along the street, and now and then the bright face of a little child rose above the horizon of a family pew. Miss Harriet Pyne lived on in the large white house, which gained more and more distinction because it suffered no changes, save successive repaintings and a new railing about its stately roof. Miss Harriet herself had moved far beyond the uncertainties of an anxious youth. She had long ago made all her decisions, and settled all necessary questions; her scheme of life was as faultless as the miniature landscape of a Japanese garden, and as easily kept in order. The only important change she would ever be capable of making was the final change to another and a better world; and for that nature itself would gently provide, and her own innocent life.

Hardly any great social event had ruffled the easy current of life since Helena Vernon's marriage. To this Miss Pyne had gone, stately in appearance and carrying gifts of some old family silver which bore the Vernon crest, but not without some protest in her heart against the uncertainties of married life. Helena was so equal to a happy independence and even to the assistance of other lives grown strangely dependent upon her quick sympathies and instinctive decisions, that it was hard to let her sink her personality in the affairs of another. Yet a brilliant English match was not without its attractions to an old-fashioned gentlewoman like Miss Pyne, and Helena herself was amazingly happy; one day there had come a letter to Ashford, in which her very heart seemed to beat with love and self-forgetfulness, to tell cousin Harriet of such new happiness and high hope. 'Tell Martha all that I say about my dear Jack,' wrote the eager girl; 'please show my letter to Martha, and tell her that I shall come home next summer and bring the handsomest and best man in the world to Ashford. I have told him all about the dear house and the dear garden; there never was such a lad to reach for cherries with his six-foot-two.' Miss Pyne, wondering a little, gave the letter to Martha, who took it deliberately and as if she wondered too, and went away to read it slowly by herself. Martha cried over it, and felt a strange sense of loss and pain; it hurt her heart a little to read about the cherry-picking. Her idol seemed to be less her own since she had become the idol of a stranger. She never had taken such a letter in her hands before, but love at last prevailed, since Miss Helena was happy, and she kissed the last page where her name was written, feeling overbold, and laid the envelope on Miss Pyne's secretary without a word.

The most generous love cannot but long for reassurance, and Martha had the joy of being remembered. She was not forgotten when the day of the wedding drew near, but she never knew that Miss Helena had asked if cousin Harriet would not bring Martha to town; she should like to have Martha there to see her married. 'She would help about the flowers,' wrote the happy girl; 'I know she will like to come, and I'll ask mamma to plan to have some one take her all about Boston and make her have a pleasant time after the hurry of the great day is over.'

Cousin Harriet thought it was very kind and exactly like Helena, but Martha would be out of her element; it was most imprudent and girlish to have thought of such a thing. Helena's mother would be far from wishing for any unnecessary guest just then, in the busiest part of her household, and it was best not to speak of the invitation. Some day Martha should go to Boston if she did well, but not now. Helena did not forget to ask if Martha had come, and was astonished by the indifference of the answer. It was the first thing which reminded her that she was not a fairy princess having everything her own way in that last day before the wedding. She knew that Martha would have loved to be near, for she could not help understanding in that moment of her own happiness the love that was hidden in another heart. Next day this happy young princess, the bride, cut a piece of a great cake and put it into a pretty box that had held one of her wedding presents. With eager voices calling her, and all her friends about her, and her mother's face growing more and more wistful at the thought of parting, she still lingered and ran to take one or two trifles from her dressing-table, a little mirror and some tiny scissors that Martha would remember, and one of the pretty handkerchiefs marked with her maiden name. These she put in the box too; it was half a girlish freak and fancy, but she could not help trying to share her happiness, and Martha's life was so plain and dull. She whispered a message, and put the little package into cousin Harriet's hand for Martha as she said good-by. She was very fond of cousin Harriet. She smiled with a gleam of her old fun; Martha's puzzled look and tall awkward figure seemed to stand suddenly before her eyes, as she promised to come again to Ashford. Impatient voices called to Helena, her lover was at the door, and she hurried away, leaving her old home and her girlhood gladly. If she had only known it, as she kissed cousin Harriet good-by, they were never going to see each other again until they were old women. The first step that she took out of her father's house that day, married, and full of hope and joy, was a step that led her away from the green elms of Boston Common* and away from her own country and those she loved best, to a brilliant, much-varied foreign life, and to nearly all the sorrows and nearly all the joys that the heart of one woman could hold or know.

On Sunday afternoons Martha used to sit by the window in Ashford and hold the wooden box which a favorite young brother, who afterward died at sea, had made for her, and she used to take out of it the pretty little box with a gilded cover that had held the piece of wedding-cake, and the small scissors, and the blurred bit of a mirror in its silver case; as for the handkerchief with the narrow lace edge, once in two or three years she sprinkled it as if it were a flower, and spread it out in the sun on the old bleaching-green, and sat near by in the shrubbery to watch lest some bold robin or cherry-bird should seize it and fly away.

IV

MISS HARRIET PYNE was often congratulated upon the good fortune of having such a helper and friend as Martha. As time went on this tall, gaunt woman, always thin, always slow, gained a dignity of behavior and simple affectionateness of look which suited the charm and dignity of the ancient house. She was unconsciously beautiful like a saint, like the picturesqueness of a lonely tree which lives to shelter unnumbered lives and to stand quietly in its place. There was such rustic homeliness and constancy belonging to her, such beautiful powers of apprehension, such reticence, such gentleness for those who were troubled or sick; all these gifts and graces Martha hid in her heart. She never joined the church because she thought she was not good enough, but life was such a passion and happiness of service that it was impossible not to be devout, and she was always in her humble place on Sundays, in the back pew next the door. She had been educated by a remembrance; Helena's young eyes forever looked at her reassuringly from a gay girlish face. Helena's sweet patience in teaching her own awkwardness could never be forgotten.

'I owe everything to Miss Helena,' said Martha, half aloud, as she sat alone by the window; she had said it to herself a thousand times. When she looked in the little keepsake mirror she always hoped to see some faint reflection of Helena Vernon, but there was only her own brown old New England face to look back at her wonderingly.

Miss Pyne went less and less often to pay visits to her friends in Boston; there were very few friends left to come to Ashford and make long visits in the summer, and life grew more and more monotonous. Now and then there came news from across the sea and messages of remembrance, letters that were closely written on thin sheets of paper, and that spoke of lords and ladies, of great journeys, of the death of little children and the proud successes of boys at school, of the wedding of Helena Dysart's only daughter; but even that had happened years ago. These things seemed far away and vague, as if they belonged to a story and not to life itself; the true links with the past were quite different. There was the unvarying flock of ground-sparrows that Helena had begun to feed; every morning Martha scattered crumbs for them from the side doorsteps while Miss Pyne watched from the dining-room window, and they were counted and cherished year by year.

Miss Pyne herself had many fixed habits, but little ideality or imagination, and so at last it was Martha who took thought for her mistress, and gave freedom to her own good taste. After a while, without any one's observing the change, the every-day ways of doing things in the house came to be the stately ways that had once belonged only to the entertainment of guests. Happily both mistress and maid seized all possible chances for hospitality, yet Miss Harriet nearly always sat alone at her exquisitely served table with its fresh flowers, and the beautiful old china which Martha handled so lovingly that there was no good excuse for keeping it hidden on closet shelves. Every year when the old cherry-trees were in fruit, Martha carried the round white old English dish with a fretwork edge, full of pointed green leaves and scarlet cherries, to the minister, and his wife never quite understood why every year he blushed and looked so conscious of the pleasure, and thanked Martha as if he had received a very particular attention. There was no pretty suggestion toward the pursuit of the fine art of housekeeping in Martha's limited acquaintance with newspapers that she did not adopt; there was no refined old custom of the Pyne housekeeping that she consented to let go. And every day, as she had promised, she thought of Miss Helena,—oh, many times in every day: whether this thing would please her, or that be likely to fall in with her fancy or ideas of fitness. As far as was possible the rare news that reached Ashford through an occasional letter or the talk of guests was made part of Martha's own life, the history of her own heart. A worn old geography often stood open at the map of Europe on the light-stand in her room, and a little old-fashioned gilt button, set with a bit of glass like a ruby, that had broken and fallen from the trimming of one of Helena's dresses, was used to mark the city of her dwelling-place. In the changes of a diplomatic life Martha followed her lady all about the map. Sometimes the button was at Paris, and sometimes at Madrid; once, to her great anxiety, it remained long at St Petersburg. For such a slow scholar Martha was not unlearned at last, since everything about life in these foreign towns was of interest to her faithful heart. She satisfied her own mind as she threw crumbs to the tame sparrows; it was all part of the same thing and for the same affectionate reasons.

V

ONE Sunday afternoon in early summer Miss Harriet Pyne came hurrying along the entry that led to Martha's room and called two or three times before its inhabitant could reach the door. Miss Harriet looked unusually cheerful and excited, and she held something in her hand. 'Where are you, Martha?' she called again. 'Come quick, I have something to tell you!'

'Here I am, Miss Pyne,' said Martha, who had only stopped to put her precious box in the drawer, and to shut the geography.

'Who do you think is coming this very night at half-past six? We must have everything as nice as we can; I must see Hannah at once. Do you remember my cousin Helena who has lived abroad so long? Miss Helena Vernon,—the Honorable Mrs Dysart, she is now.'

'Yes, I remember her,' answered Martha, turning a little pale.

'I knew that she was in this country, and I had written to ask her to come for a long visit,' continued Miss Harriet, who did not often explain things, even to Martha, though she was always conscientious about the kind messages that were sent back by grateful guests. 'She telegraphs that she means to anticipate her visit by a few days and come to me at once. The heat is beginning in town, I suppose. I daresay, having been a foreigner so long, she does not mind traveling on Sunday. Do you think Hannah will be prepared? We must have tea a little later.'

'Yes, Miss Harriet,' said Martha. She wondered that she could speak as usual, there was such a ringing in her ears. 'I shall have time to pick some fresh strawberries; Miss Helena is so fond of our strawberries.'

'Why, I had forgotten,' said Miss Pyne, a little puzzled by something quite unusual in Martha's face. 'We must expect to find Mrs Dysart a good deal changed, Martha; it is a great many years since she was here; I have not seen her since her wedding, and she has had a great deal of trouble, poor girl. You had better open the parlor chamber, and make it ready before you go down.'

'It is all ready,' said Martha. 'I can carry some of those little sweet-brier roses upstairs before she comes.'

'Yes, you are always thoughtful,' said Miss Pyne, with unwonted feeling.

Martha did not answer. She glanced at the telegram wistfully. She had never really suspected before that Miss Pyne knew nothing of the love that had been in her heart all these years; it was half a pain and half a golden joy to keep such a secret; she could hardly bear this moment of surprise.

Presently the news gave wings to her willing feet. When Hannah, the cook, who never had known Miss Helena, went to the parlor an hour later on some errand to her old mistress, she discovered that this stranger guest must be a very important person. She had never seen the tea-table look exactly as it did that night, and in the parlor itself there were fresh blossoming boughs in the old East India jars, and lilies in the paneled hall, and flowers everywhere, as if there were some high festivity.

Miss Pyne sat by the window watching, in her best dress, looking stately and calm; she seldom went out now, and it was almost time for the carriage. Martha was just coming in from the garden with the strawberries, and with more flowers in her apron. It was a bright cool evening in June, the golden robins sang in the elms, and the sun was going down behind the apple-trees at the foot of the garden. The beautiful old house stood wide open to the long-expected guest.

'I think that I shall go down to the gate,' said Miss Pyne, looking at Martha for approval, and Martha nodded and they went together slowly down the broad front walk.

There was a sound of horses and wheels on the roadside turf: Martha could not see at first; she stood back inside the gate behind the white lilac-bushes as the carriage came. Miss Pyne was there; she was holding out both arms and taking a tired, bent little figure in black to her heart. 'Oh, my Miss Helena is an old woman like me!' and Martha gave a pitiful sob; she had never dreamed it would be like this; this was the one thing she could not bear.

'Where are you, Martha?' called Miss Pyne. 'Martha will bring these in; you have not forgotten my good Martha, Helena?' Then Mrs Dysart looked up and smiled just as she used to smile in the old days. The young eyes were there still in the changed face, and Miss Helena had come.

That night Martha waited in her lady's room just as she used, humble and silent, and went through with the old unforgotten loving services. The long years seemed like days. At last she lingered a moment trying to think of something else that might be done, then she was going silently away, but Helena called her back. She suddenly knew the whole story and could hardly speak.

'Oh, my dear Martha!' she cried, 'won't you kiss me good-night? Oh, Martha, have you remembered like this, all these long years!'

BOLD WORDS AT THE BRIDGE

I

' "WELL, now," says I, "Mrs Con'ly," says I, "how ever you may tark, 't is nobody's business and I wanting to plant a few pumpkins for me cow in among me cabbages. I've got the right to plant whatever I may choose, if it's the divil of a crop of t'istles in the middle of me ground." "No ma'am, you ain't," says Biddy Con'ly; "you ain't got anny right to plant t'istles that's not for the public good," says she; and I being so hasty wit' me timper, I shuk me fist in her face then, and herself shuk her fist at me. Just then Father Brady come by, as luck ardered, an' recommended us would we keep the peace. He knew well I'd had my provocation; 't was to herself he spoke first. You'd think she owned the whole corporation. I wished I'd t'rown her over into the wather, so I did, before he come by at all. 'T was on the bridge the two of us were. I was stepping home by meself very quiet in the afthernoon to put me tay-kittle on for supper, and herself overtook me,—ain't she the bold thing!

' "How are you the day, Mrs Dunl'avy?" says she, so mincin' an' preenin', and I knew well she'd put her mind on having words wit' me from that minute. I'm one that likes to have peace in the neighborhood, if it wa'n't for the likes of her, that makes the top of me head lift and clat'* wit' rage like a pot-lid!'

'What was the matter with the two of you?' asked a listener, with simple interest.

'Faix* indeed, 't was herself had a thrifle of melons planted the other side of the fince,' acknowledged Mrs Dunleavy. 'She said the pumpkins would be the ruin of them intirely. I says, and 't was thrue for me, that I'd me pumpkins planted the week before she'd dropped anny old melon seed into the ground, and the same bein' already dwining* from so manny bugs. Oh, but she's blackhearted to give me the lie about it, and say those poor things was all up, and she'd thrown lime on 'em to keep away their inemies when she first see me come out betune me cabbage rows. How well she knew what I might be doing! Me cabbages grows far apart and I'd plinty of room, and if a pumpkin vine gets attention you can entice it wherever you pl'ase and it'll grow fine and long, while the poor cabbages ates* and grows fat and round, and no harm to annybody, but she must pick a quarrel with a quiet 'oman in the face of every one.

'We were on the bridge, don't you see, and plinty was passing by with their grins, and loitering and stopping afther they were behind her back to hear what was going on betune us. Annybody does be liking to get the sound of loud talk an' they having nothing better to do. Biddy Con'ly, seeing she was well watched, got the airs of a pr'acher, and set down whatever she might happen to be carrying and tried would she get the better of me for the sake of their admiration. Oh, but wa'n't she all drabbled and wet from the roads, and the world knows meself for a very tidy walker!

' "Clane the mud from your shoes if you're going to dance;" 't was all I said to her, and she being that mad she did be stepping up and down like an old turkey-hin, and shaking her fist all the time at me. "Coom now, Biddy," says I, "what put you out so?" says I. "Sure, it creeps me skin when I looks at you! Is the pig dead," says I, "or anny little thing happened to you, ma'am? Sure this is far beyond the rights of a few pumpkin seeds that has just cleared the ground!" and all the folks laughed. I'd no call to have tark with Biddy Con'ly before them idle b'ys and gerrls, nor to let the two of us become their laughing-stock. I tuk up me basket, being ashamed then, and I meant to go away, mad as I was. "Coom, Mrs Con'ly!" says I, "let bygones be bygones; what's all this whillalu* we're afther having about nothing?" says I very pleasant.

' "May the divil fly away with you, Mary Dunl'avy!" says she then, "spoiling me garden ground, as every one can see, and full of your bold talk. I'll let me hens out into it this afternoon, so I will," says she, and a good deal more. "Hold off," says I, "and remember what fell to your aunt one day when she sint her hins in to pick a neighbor's piece, and while her own back was turned they all come home and had every

sprouted bean and potatie heeled out in the hot sun, and all
her fine lettuces picked into Irish lace. We've lived neighbors," says I, "thirteen years," says I; "and we've often had
words together above the fince," says I, "but we're neighbors
yet, and we've no call to stand here in such spectacles and
disgracing ourselves and each other. Coom, Biddy," says I,
again, going away with me basket and remembering Father
Brady's caution whin it was too late. Some o' the b'ys went off
too, thinkin' 't was all done.

' "I don't want anny o' your Coom Biddy's," says she, stepping at me, with a black stripe across her face, she was that
destroyed with rage, and I stepped back and held up me
basket between us, she being bigger than I, and I getting no
chance, and herself slipped and fell, and her nose got a clout
with the hard edge of the basket, it would trouble the saints to
say how, and then I picked her up and wint home with her to
thry and quinch the blood. Sure I was sorry for the crathur an'
she having such a timper boiling in her heart.

' "Look at you now, Mrs Con'ly," says I, kind of soft, "you
'ont be fit for mass these two Sundays with a black eye like this,
and your face arl scratched, and every bliguard has gone the
lingth of the town to tell tales of us. I'm a quiet 'oman," says
I, "and I don't thank you," says I, whin the blood was
stopped,—"no, I don't thank you for disgracin' an old neighbor like me. 'T is of our prayers and the grave we should be
thinkin', and not be having bold words on the bridge."
Wisha!* but I t'ought I was after spaking very quiet, and up she
got and caught up the basket, and I dodged it by good luck,
but after that I walked off and left her to satisfy her foolishness
with b'ating the wall if it pl'ased her. I'd no call for her
company anny more, and I took a vow I'd never spake a word
to her again while the world stood. So all is over since then
betune Biddy Con'ly and me. No, I don't look at her at all!'

II

SOME time afterward, in late summer, Mrs Dunleavy stood,
large and noisy, but generous-hearted, addressing some remarks from her front doorway to a goat on the sidewalk. He

was pulling some of her cherished foxgloves through the picket fence, and eagerly devouring their flowery stalks.

'How well you rache through an honest fince, you black pirate!' she shouted; but finding that harsh words had no effect, she took a convenient broom, and advanced to strike a gallant blow upon the creature's back. This had the simple effect of making him step a little to one side and modestly begin to nibble at a tuft of grass.

'Well, if I ain't plagued!' said Mrs Dunleavy sorrowfully; 'if I ain't throubled with every wild baste, and me cow that was some use gone dry very unexpected, and a neighbor that's worse than none at all. I've nobody to have an honest word with, and the morning being so fine and pleasant. Faix, I'd move away from it, if there was anny place I'd enjoy better. I've no heart except for me garden, me poor little crops is doing so well; thanks be to God, me cabbages is very fine. There does be those that overlooked me pumpkins for the poor cow; they're no size at all wit' so much rain.'

The two small white houses stood close together, with their little gardens behind them. The road was just in front, and led down to a stone bridge which crossed the river to the busy manufacturing village beyond. The air was fresh and cool at that early hour, the wind had changed after a season of dry, hot weather; it was just the morning for a good bit of gossip with a neighbor, but summer was almost done, and the friends were not reconciled. Their respective acquaintances had grown tired of hearing the story of the quarrel, and the novelty of such a pleasing excitement had long been over. Mrs Connelly was thumping away at a handful of belated ironing, and Mrs Dunleavy, estranged and solitary, sighed as she listened to the iron. She was sociable by nature, and she had an impulse to go in and sit down as she used at the end of the ironing table.

'Wisha, the poor thing is mad at me yet, I know that from the sounds of her iron; 't was a shame for her to go picking a quarrel with the likes of me,' and Mrs Dunleavy sighed heavily and stepped down into her flower-plot to pull the distressed foxgloves back into their places inside the fence. The seed had been sent her from the old country, and this was the first year they had come into full bloom. She had been hoping that the sight of them would melt Mrs Connelly's heart into some expression of friendliness, since they had come from adjoining parishes in old County Kerry.* The goat lifted his head, and gazed at his enemy with mild interest; he was pasturing now by the roadside, and the foxgloves had proved bitter in his mouth.

Mrs Dunleavy stood looking at him over the fence, glad of even a goat's company.

'Go 'long there; see that fine little tuft ahead now,' she advised him, forgetful of his depredations. 'Oh, to think I've nobody to spake to, the day!'

At that moment a woman came in sight round the turn of the road. She was a stranger, a fellow country-woman, and she carried a large newspaper bundle and a heavy handbag. Mrs Dunleavy stepped out of the flower-bed toward the gate, and waited there until the stranger came up and stopped to ask a question.

'Ann Bogan don't live here, do she?'

'She don't,' answered the mistress of the house, with dignity.

'I t'ought she didn't; you don't know where she lives, do you?'

'I don't,' said Mrs Dunleavy.

'I don't know ayther; niver mind, I'll find her; 't is a fine day, ma'am.'

Mrs Dunleavy could hardly bear to let the stranger go away. She watched her far down the hill toward the bridge before she turned to go into the house. She seated herself by the side window next Mrs Connelly's, and gave herself to her thoughts. The sound of the flatiron had stopped when the traveler came to the gate, and it had not begun again. Mrs Connelly had gone to her front door; the hem of her calico dress could be plainly seen, and the bulge of her apron, and she was watching the stranger quite out of sight. She even came out to the doorstep, and for the first time in many weeks looked with friendly intent toward her neighbor's house. Then she also came and sat down at her side window. Mrs Dunleavy's heart began to leap with excitement.

'Bad cess* to her foolishness, she does be afther wanting to come round; I'll not make it too aisy for her,' said Mrs Dunleavy, seizing a piece of sewing and forbearing to look up. 'I don't know who Ann Bogan is, annyway; perhaps herself does, having lived in it five or six years longer than me. Perhaps she knew this woman by her looks, and the heart is out of her with wanting to know what she asked from me. She can sit there, then, and let her irons grow cold!

'There was Bogans living down by the brick mill when I first come here, neighbors to Flaherty's folks,' continued Mrs Dunleavy, more and more aggrieved. 'Biddy Con'ly ought to know the Flahertys, they being her cousins. 'T was a fine loud-talking 'oman; sure Biddy might well enough have heard her inquiring of me, and have stepped out, and said if she knew Ann Bogan, and satisfied a poor stranger that was hunting the town over. No, I don't know anny one in the name of Ann Bogan, so I don't,' said Mrs Dunleavy aloud, 'and there's nobody I can ask a civil question, with every one that ought to be me neighbors stopping their mouths, and keeping black grudges whin 't was meself got all the offince.'

'Faix 't was meself got the whack on me nose,' responded Mrs Connelly quite unexpectedly. She was looking squarely at the window where Mrs Dunleavy sat behind the screen of blue mosquito netting. They were both conscious that Mrs Connelly made a definite overture of peace.

'That one was a very civil-spoken 'oman that passed by just now,' announced Mrs Dunleavy, handsomely waiving the subject of the quarrel and coming frankly to the subject of present interest. 'Faix, 't is a poor day for Ann Bogans; she'll find that out before she gets far in the place.'

'Ann Bogans was plinty here once, then, God rest them! There was two Ann Bogans, mother and daughter, lived down by Flaherty's when I first come here. They died in the one year, too; 't is most thirty years ago,' said Bridget Connelly, in her most friendly tone.

' "I'll find her," says the poor 'oman as if she'd only to look; indeed, she's got the boldness,' reported Mary Dunleavy, peace being fully restored.

' 'T was to Flaherty's she'd go first, and they all moved to La'rence twelve years ago, and all she'll get from anny one would be the address of the cimet'ry. There was plenty here knowing to Ann Bogan once. That 'oman is one I've seen long ago, but I can't name her yet. Did she say who she was?' asked the neighbor.

'She didn't; I'm sorry for the poor 'oman, too,' continued Mrs Dunleavy, in the same spirit of friendliness. 'She'd the expectin' look of one who came hoping to make a nice visit and find friends, and herself lugging a fine bundle. She'd the looks as if she'd lately come out;* very decent, but old-fashioned. Her bonnet was made at home annyways, did ye mind? I'll lay it was bought in Cork when it was new, or maybe 't was from a good shop in Bantry or Kinmare,* or some o' those old places. If she'd seemed satisfied to wait, I'd made her the offer of a cup of tay, but off she wint with great courage.'

'I don't know but I'll slip on me bonnet in the afthernoon and go find her,' said Biddy Connelly, with hospitable warmth. 'I've seen her before, perhaps 't was long whiles ago at home.'

'Indeed I thought of it myself,' said Mrs Dunleavy, with approval. 'We'd best wait, perhaps, till she'd be coming back; there's no train now till three o'clock. She might stop here till the five, and we'll find out all about her. She'll have a very lonesome day, whoiver she is. Did you see that old goat 'ating the best of me fairy-fingers that all bloomed the day?' she asked eagerly, afraid that the conversation might come to an end at any moment; but Mrs Connelly took no notice of so trivial a subject.

'Me melons is all getting ripe,' she announced, with an air of satisfaction. 'There's a big one must be ate now while we can; it's down in the cellar cooling itself, an' I'd like to be dropping it, getting down the stairs. 'T was afther picking it I was before breakfast, itself having begun to crack open. Himself was the b'y that loved a melon, an' I ain't got the heart to look at it alone. Coom over, will ye, Mary?'

''Deed then an' I will,' said Mrs Dunleavy, whose face was close against the mosquito netting. 'Them old pumpkin vines was no good anny way; did you see how one of them had the invintion, and wint away up on the fince entirely wit' its great flowers, an' there come a rain on 'em, and so they all blighted? I'd no call to grow such stramming* great things in my piece annyway, 'ating up all the goodness from me beautiful cabbages.'

III

THAT afternoon the reunited friends sat banqueting together and keeping an eye on the road. They had so much to talk over and found each other so agreeable that it was impossible to dwell with much regret upon the long estrangement. When the melon was only half finished the stranger of the morning, with her large unopened bundle and the heavy handbag, was seen making her way up the hill. She wore such a weary and disappointed look that she was accosted and invited in by both the women, and being proved by Mrs Connelly to be an old acquaintance, she joined them at their feast.

'Yes, I was here seventeen years ago for the last time,' she explained. 'I was working in Lawrence, and I came over and spent a fortnight with Honora Flaherty; then I wint home that year to mind me old mother, and she lived to past ninety. I'd nothing to keep me then, and I was always homesick afther America, so back I come to it, but all me old frinds and neighbors is changed and gone. Faix, this is the first welcome I've got yet from anny one. 'T is a beautiful welcome, too,— I'll get me apron out of me bundle, by your l'ave, Mrs Con'ly. You've a strong resemblance to Flaherty's folks, dear, being cousins. Well, 't is a fine thing to have good neighbors. You an' Mrs Dunleavy is very pleasant here so close together.'

'Well, we does be having a hasty word now and then, ma'am,' confessed Mrs Dunleavy, 'but ourselves is good neighbors this manny years. Whin a quarrel's about nothing betune friends, it don't count for much, so it don't.'

'Most quarrels is the same way,' said the stranger, who did not like melons, but accepted a cup of hot tea. 'Sure, it always takes two to make a quarrel, and but one to end it; that's what me mother always told me, that never gave anny one a cross word in her life.'

' 'T is a beautiful melon,' repeated Mrs Dunleavy for the

seventh time. 'Sure, I'll plant a few seed myself next year; me pumpkins is no good afther all me foolish pride wit' 'em. Maybe the land don't suit 'em, but glory be to God, me cabbages is the size of the house, an' you'll git the pick of the best, Mrs Con'ly.'

'What's melons betune friends, or cabbages ayther, that they should ever make any trouble?' answered Mrs Connelly handsomely, and the great feud was forever ended.

But the stranger, innocent that she was the harbinger of peace, could hardly understand why Bridget Connelly insisted upon her staying all night and talking over old times, and why the two women put on their bonnets and walked, one on either hand, to see the town with her that evening. As they crossed the bridge they looked at each other shyly, and then began to laugh.

'Well, I missed it the most on Sundays going all alone to mass,' confessed Mary Dunleavy. 'I'm glad there's no one here seeing us go over, so I am.'

' 'T was ourselves had bold words at the bridge, once, that we've got the laugh about now,' explained Mrs Connelly politely to the stranger.

THE HONEY TREE

I

THERE was a great piece of news in Hillborough Friday night, which was told at a friendly meeting in Joel Simmons's store. It was the first autumn evening when the air felt frosty enough for a fire; the outside benches along the store front were wholly deserted for the first time that season. The newspapers reached Hillborough in the morning, so that those few citizens who took the *Tribune* or the *Herald* had time enough to digest any information received before the evening gathering, and to impart their refreshed ideas to those persons who came down from the upper hill country after supper.

The Reverend Mr Dennett was the last to arrive and to ask for his belated morning mail; he had been away all day at the funeral of a former parishioner, and some of the men who sat in the store inquired about the day's events, and possessed themselves of whatever interesting facts might be available. The minister was always pleasantly communicative.

'What has been going on here to-day?' He turned back to put the question just as he reached the door; there was some influence or sudden instinct of sympathy which impelled him; perhaps he noticed an unusual eagerness in his parishioners' faces.

'Not much o' anything 'bout here,' answered old Captain Foss before anybody else could speak. 'No, sir, I don't know o' anything special this part o' the parish, but somewheres up there by Sunday Mountain there's b'en a bee tree discovered; one o' old Mis' Prime's gran'children, the Hopper boy, found it, and they say there's like to be fifty or sixty pound o' this year's new honey. Asy Hopper himself informed Martin Wells as he was ridin' by, this evenin'. I've got the facts right, 'ain't I, Martin?' inquired the old man, politely giving up the floor now that he had possessed himself of the glory of telling the news to the minister.

'Why, that *is* news!' exclaimed Mr Dennett. 'Fifty pounds of

honey is indeed a valuable acquisition. I suppose that
Mr Hopper intends to market some of it. I shall be glad
to patronize him myself; honey is very soothing to the preach-
er's throat. Yes, I should feel personally grateful of the
opportunity.'

This ecclesiastical tribute to the efforts of the hive seemed
to put the occasion on a still higher and more interesting
level. One man after another said that he would be willing to
put his name down for five or six pounds, but an eager young
voice interrupted these calm appropriations.

'His father says Johnny Hopper's goin' to have all the
honey hisself to do what he's a mind to with. 'T was Johnny
found the tree,' piped the boy, with such a displeasing importance in his way of giving information that even the minister's
face fell a little.

'Sho! sho!'* said the Captain, ready with instant rebuke.
'His father said Johnny should have all the honey that was
good for him, I guess. 'Tis too large a quantity to eat all up at
home, and they ain't very well off neither.'

'I ain't goin' to take none if a pack o' boys has been pawin'
into it!' proclaimed the storekeeper, excitedly. 'I ain't goin'
to have what I expect to dispose of fetched back here to the
store 'count o' bein' full o' dry bark an' pine spills, dead bees,
an' all them sorts o' trollick.* I guess Asy Hopper'll know
enough to smoke out them bees, and wedge the tree right
open, and get that honey out proper, 's he knows we should
want it. I guess he won't use Johnny no way but right neither,'
he added, being a kind-hearted man, and seeing the look of
dismay on the young speaker's face.

'No, 't was Johnny found the tree,' repeated the school-
teacher, a long-faced man, 'but he will want to do just what his
father considers best, like a real good son.' Mr Dunn sighed
heavily as he finished this charge to the elect, and rose from
an uneasy crate where he had been sitting, and took his
dignified evening way toward the door. Bill Phillips, the inelegant boy toward whom his utterance was directed, put out
his tongue as far as it would go, and was red in the face from
the protracted effort before the door was shut behind his
natural enemy. The minister looked grave and disapproving,
but some of the other men laughed.

'You don't want to be sassy like that!' said the old Captain to Bill. 'You'd get hove right overboard if they see you do that on a marchant vessel, sir, now I tell you! You was talking to me about follerin' the sea t' other day,' he ended, severely, but with such kindness and sincere interest that the lad looked abashed, and presently sidled off among the barrels and gained his unnoticed liberty.

'Some folks is said to be deadly p'isoned if they trifle with honey,' announced old Mr Jenkins, warningly, from one of the arm-chairs. 'I don't know 's it's very common to hear o' such cases, but my mother had an aunt by marr'ge that was throwed into complete fits. They thought 't was her own notion an' she'd heared o' somebody else that was affected so, or suthin', but they tried her four or five times puttin' honey onbeknownst into sweet-cake or the like o' that, and she'd be right into them fits without fail, sir! After the last time she come out on 'em feelin' kind o' slim for a good while, and so they didn't tax her no more. They thought if they could once get a good portion consumed, and she was none the wuss, they'd laugh her out of it. She was al'ays a notional person.'

'Better leave good honest honey for such as desires it,' growled Martin Wells. 'Asy told me hisself they should have a plenty to winter 'em, and like 's not some to spare. He promised me what I could use, anyway.'

Every eye in the company glistened at this information, and there was a silence, as if to resolve upon a course that could be properly maintained. It was a great many years since honey had been plentiful in Hillborough, and neither Bill Phillips nor Johnny Hopper felt a deeper interest in the simple luxury than these elderly men.

'Goes good on a slice o' rye bread and butter,' said Captain Foss, smacking his lips. 'Wife 'n' me used to carry a little crock along in our seafarin' days; honey or stewed cranberries was our gre't treat for Sunday night supper aboard; an', Lord! how we used to mourn 'em when they was all gone, an' we'd got three months afore us sometimes ere we'd make our port! Dried apples, even them'd get mouldy, and down we'd come at last to plain hardtack an' beef out o' the old harness-cask.'*

The storekeeper gave a reassuring glance at his shelves, which were bending with canned goods and vegetables and bright California fruits. 'I could fit you and Mis' Foss out very handsome now to go right round the world,' he announced. But the Captain sniffed, and worked the ferrule of his cane back and forth angrily in a familiar crack of the floor.

'Them things!' he exclaimed. 'I'd starve fust, and so would Mis' Foss. They taste all of 'em alike; they'd give some folks onnecessary fits worse 'n them we've heared described. Them cans would all bulge their tops and go off like guns agin the upper deck, take 'em into some o' the latitudes o' heat where I've been.' And the storekeeper was humbled to the earth.

'You had some o' my peaches for supper last night, anyways,' he ventured. 'Mis' Foss sent over in a hurry for 'em, sayin' she'd got onexpected comp'ny come.'

'I observed her preserves was dreadful poor for once,' glared the Captain. 'Oh, well, sir, I ain't disputin' nor cryin' down your business. They seem shiftless to me, these new-fangled notions o' eatin', but then I be an old sailor,' and he laughed a little. 'Dare say I should be glad enough on 'em, come to go to sea agin an' be short o' stores!' It was an irresistible chuckle, was the Captain's; and cheerfulness was at once restored.

The minister, who had been patiently waiting for a pound of tea to be weighed and put up, now said good-night and went away.

'Poor creatur'! I guess he knows whether canned goods is nourishing or not,' said Martin Wells, impulsively. 'They've got a story up our way that poor Mis' Dennett ain't no gre't of a housekeeper; my woman is dreadful 'tached to the minister since he was so feelin' for her the time we lost our little girl, an' she can't let me ride down here to the Plains 'thout a loaf o' her good bread, or a pie, or somethin' for 'em.'

'I expect she's had it hard, his wife has, with their large family. Their minds is turned other ways, ministers' folks is,' commented the storekeeper, compassionately. 'I'll bate* you some o' that honey 'll get to the pa'sonage, and if Mis' Wells's extra bread puts into port same day, they'll have a treat, sure 's can be!'

Martin Wells blushed with inward delight at this tribute.

There was a man, John Timms by name, who had not

spoken. He was very deaf, and had waited till the talk was done before he put a modest question.

'What'd you say when you fust come in? I didn't catch the drift on 't,' he asked, as the old Captain rose to go home, and the others knew by this signal that the evening was over.

'I said that Hopper's folks had found a bee tree up side o' Sunday Mountain,' said Martin, bawling into his ear.

'Much honey in it?' asked Timms in a stifled voice.

'Fifty or sixty pound; this year's make!'

'Guess they'll be havin' plenty o' company up to Hopper's if this good weather holds,' prophesied the latest receiver of the happy news.

II

'MOTHER, you ain't thinkin' o' goin' 'way up there side o' the mountain!' exclaimed Mrs Hopper, Johnny's mother, next morning. She was busy getting out all the large dishes from her cupboard, and had already brought some large clean basswood chopping-trays and bowls from the outer store-room. Grandma Prime made her appearance dressed for the outer air, and had her big umbrella in hand as if she would need a staff. 'They said 't was nigh a mile off where they found the tree,' protested the younger woman, anxiously. ' 'Tis rough under foot; there, you might catch your foot in a root, and get a fall you wouldn't be better of all winter long!'

'Ann Sarah, I've clim' Sunday Mountain before ever you was born, an' if anybody feels to do a thing they *can* do it; my mind is set on gettin' up to see that bee tree Johnny found; an' I'm a-goin'. I'll take it slow. If you keep a-don'tin' me an' makin' me feel I'm past everything, my heart will break. I've al'ays been used to my liberty,' and her old face quivered.

'Why, of course you can go, you dear creatur',' said Ann Sarah, hastily trying to make amends. 'Take it slow, as you say, mother; we'll work along together. I don't know where that Johnny is, for my part; he said he'd go up with 's father and Bill Phillips an' the rest o' the boys an' men, and show 'em where 't was, and then he'd come right back and help me with these bowls and buckets and things. Mis' Wells come along with Martin whilst you took your nap, an' said she was goin' up to see 'em fight the bees an' get the honey out; she never see such a sight in her life where she come from. It ain't but one o'clock now; they must ha' got their dinner out o' the way 'arlier 'n we did ours.'

'I didn't stop to take no gre't of a nap, for all we had such a drivin' mornin',' said Grandma Prime, with importance. 'I heared voices, and I wanted to be off, myself. Well, 't is a lovely afternoon, an' happens just right to have it come a Saturday!'

'I declare you're pleased as a girl, mother,' said Mrs Hopper, proudly. 'You look well an' young as ever you did!'

'Come, let's go right along!' urged the old adventurer; ' 't will be all over before we git there! Johnny never 'll think o' desertin' the rest on 'em once the real play begins.'

'See here, I do' know but I can put a number o' these wooden things that's light right into the bushel basket, and car' an extra pail on my arm. I wish we had a stick to run through the basket handles and take it right between us,' said Ann Sarah.

'Run this umbrella through,' directed Grandma Prime. 'Here, I sha'n't require it. You'll waste an hour longer huntin' for somethin' else!' And they started together up the wood-road like a careful pair of steady yokemates, with the umbrella fast held between them.

'I expect they'll have them bees all coped with, and be wonderin' what they've got to put their honey in, and be ter'ble glad to see us a-comin',' said grandma, stopping on the steep hill-side to take breath.

III

BEFORE long they heard the blows of an axe and the loud sound of voices. The two women were more eager of heart than they were swift of foot; if the mother was hindered by age, the daughter was a stout person not given to mountaineering. It was a beautiful October afternoon; the dark woods still kept their frosty morning fragrance, but in the open spaces the sun felt as if it were still June. All the blue-jays were talking and scolding at each other; their voices were not unlike those of the bee-hunters themselves, who may have been disturbing them.

'Yes, I hear our folks now very plain,' said Grandma Prime, whose ears were not quite so keen as her daughter's. 'I hear 'em plain. Let's get along a little mite faster, if it's so you can, Ann Sarah.'

The honey tree stood at the edge of an open space of smooth turf. It was an old apple-tree, and behind it was a thick growth of young pines. These were fast covering a disused pasture, which had been burnt so dry every year in midsummer, and was so poorly watered, that Asa Hopper had let the forest in at last to take full possession. The apple-tree was a poor ungrafted seedling; its fruit was eatable only by boys; and for lack of nourishment in the thin soil, its thick short trunk had long ago grown hollow. The bees had come and gone through a large knot-hole near the ground; only a few side branches looked alive; it had long been the home of squirrels before the bees took it. There were a few knurly little yellow cider-apples on the mossy twigs.

'I can remember this tree when I was a girl,' said Grandma Prime, with much importance; 'it had dreadful pretty pink blossoms then, but the fruit was poorer than most. So 't was this tree! Why, I should have known well enough if you'd told me, Johnny.' But Johnny took no notice of what any woman might say; he was busy with a man's work, and viewed their arrival, as he had received Mrs Wells's earlier and somewhat forward advice, with great indifference. He and Bill Phillips had already suffered much disfigurement of countenance, for the smoking-out process had been most unsuccessful at first, while late-returning bees were still to be met and despatched with birch and hemlock boughs, and the fray was by no means over. There was a small fire burning, and twisted wisps of damp straw, and sulphur fuming on live coals that were heaped on a piece of bark, were still in requisition.

'You'd ought to have waited until dark to smoke 'em, or till some rainy day when they were all stopping to home,' advised Grandma Prime, with the air of an expert, after Asa Hopper had made a blind run in among the little pines with a bee about his ears; but until that wise utterance it seemed to have occurred neither to him nor to anybody else to delay the great encounter. At last the smoking process was over, the tree was cut down, they had wedged the tough trunk, and the men and boys all insisted upon giving orders together, while the women looked on as if at a splendid sight of valor.

'There she goes!' shouted Johnny at last, as the wedges and a crowbar finally prevailed, and the old tree was cleft with a loud tearing sound, and lay in two hollow halves apart, solid with honeycomb through the best part of its length. A few despoiled bees crept about bewildered in the bright slow drops that glistened where the wax had crushed or parted. Johnny and Bill Phillips, and the men too, gave a shout of triumph. There were no fifty pounds in view, but there was really more honey than they could possibly eat.

'There, come here and look, grandma!' called Johnny, returning to his old allegiance, and forgetting his manly scorn of the incompetent sex. 'Look there, grandma. What do you call that?' cried Johnny again, and stood to receive her admiration like a hero before the Athenian populace. Then he clutched at a large piece of honeycomb and took his due reward; the poor bees who had gathered it were trampled and destroyed as if they had lived but to minister to the glory and delight of others, like the vanquished army on the shore of Marathon.*

It was about four o'clock or a little earlier when the hunters started to go back to the house. The bowls and trays in the basket with which Grandma Prime and Mrs Hopper had toiled up the mountain slope were not all needed, but most of them were well filled with honey, and everybody took one to carry, even the eldest of the party, who steadied herself well enough with the umbrella. Some of the old brown comb was left behind for another day, and a good deal of new honey had leaked into the grass, but a thin, wandering bear snuffed these treasures on the light October breeze, and came that night to feast upon the honeyed ground, so that nothing was wasted.

IV

As the rich and happy company came down the wood-path and drew near the house, they saw a horse and wagon hitched to the fence, and a top-buggy beyond that, and there were several persons standing in the road by twos and threes, all looking off at the view.

'Why, what's the matter?' exclaimed Mrs Hopper. 'We oughtn't to have left the house all this time, all of us to once. I never thought to lock none o' the doors. You don't expect we've been afire? Why, see all these folks!'

'I guess they're just out strollin', 't is such a pleasant Saturday afternoon, Ann Sarah,' answered the old lady. 'Some of 'em's like to come in an' call. I wish we was dressed up better.'

'That's Joel Simmons's hoss,' said Martin Wells, innocently. 'I guess he wants to speak about engaging your fowls for Thanksgivin'.'

'I declare I do believe there's Cap'n Foss rode up to see us,' announced Grandma Prime. 'And—yes, there's Mis' Foss too, that I haven't seen up here for so long I can't remember when! I didn't know's she'd ever get up the mountain again;' and the good soul, forgetting her own weariness, hurried along to welcome a chosen friend.

The people who were walking up the road looked somewhat abashed as they hung about the door, and each of them began to make excuse. They all declared that the beautiful afternoon had tempted them out for a walk. There were seven or eight of these guests together, and they had to be coaxed before they consented to come in. Mrs Hopper looked at Martin Wells's wife with a funny little smile when she had at last prevailed over such reluctance, and Mrs Wells smiled back with comprehension and amusement.

'We've all been up on the mountain; our Johnny found a honey tree yesterday,' said Ann Sarah Hopper after she had followed them in. 'I want, now you're here, that you should all stop and have some,' she told the silent roomful as if she expected them to be surprised, and there was a feeble murmur of approval from one or two. Mrs Foss and Grandma

Prime were sitting together, holding each other's hand.
Grandma Prime looked happy but a little pale, and she still
kept hold of her small wooden bowl full of honey.

'We passed the teacher a little ways back. He's out botaniz-
ing,' said one of the young women, impulsively, now that the
first stiffness was over. 'He said he was looking for some scarce
bush that has a yellow bloom this time o' year. We asked him
to come along with us, but he said he might join us a little
later on; he was going a piece further up the road,' she added.
'We told him if 't was witch-hazel he was looking for, he
wouldn't find any quite so late.'

'Ain't it a kind of a honey-colored flower?' inquired Johnny
Hopper, smartly, with a queer brightness coming into his
eyes. He had just deposited the chopping-tray on the table
with lofty triumph, and then, as he viewed the company that
already filled every chair in the large kitchen, he cast a wistful
glance at his treasure, as if he wished it were in a safer place.

'Johnny dear,' said his mother, coming from the cupboard
with her hands full of saucers—'Johnny, they say Mr Dunn,
your school-teacher, 's down the road; you go ask him to
come right up an' have some of your nice honey, won't you?'

'*You* go, Bill,' commanded Johnny, coldly, and fled out
through the shed and up to the woods to head off his father,
who was laden with axe and crowbar and a heavy yellow bowl,
and really needed his succor. But there was a look of rueful-
ness on Johnny's face. This did not look like a winter's store
of honey for one's self, and still less like having enough to sell
besides, so that a fellow would indeed be rich. Skates and a
man's gun were rapidly disappearing down the throats of
greedy idlers, and poor Johnny's heart felt as if it were like to
break.

'I expected they'd gather by Sunday,' said his father, laughing at his son when they met. 'They *be* pretty prompt, but the
nice weather 'sort o' helped 'em. Here, you take this bowl
from me, and I'll step back and get my axe. I had to leave it
hooked on a limb back here; it wouldn't gybe* with the old
crowbar nowhow.'

'Father, ain't there some safe place up here where we can
leave the bowl?' Johnny besought him with trembling lip.

'There's a sight o' folks down to the house, an' they say there's more a-comin'.'

'Why, yes,' said his father, soberly. 'I'm glad to have somethin' to give the folks. I made out I'd let your mother have some to give away; we're goin' to have more than enough for ourselves;' and he looked down at the boy in a kindly manner.

'I want Mis' Foss and the Cap'n to have some,' said Johnny, 'and I don't mind about Mr Simmons—he's a real good man; but plague take the rest of 'em!'

'You're goin' to be just like other folks when you grow up,' remarked his father, and burst into a funny little laugh. It might have been to a small boy's disadvantage, or it might not. 'Here, you trot along to the house, an' I'll see to the big yellow bowl. Your mother'll be needing you, and I'm all honey up to my elbows. I've got to go an' wash me off, down to the brook.'

The boy obeyed, and returned just in time to see a long black coat disappear within the front door. It was the minister, and his wife was with him; they had come to make their regular parochial visitation.

The next moment Mrs Wells overtook Johnny, all out of breath with haste. 'Here, dear, you help me carry these things,' she said, carefully giving him a large brown loaf of cake. 'I thought your mother'd need a little shorin' up with such a party comin' in on her, and I just run home and brought my Saturday's baking right over,' said the kind-hearted, generous woman. She was always called the best of neighbors. There was a look of delight and social excitement in her face, which was suddenly reflected in the anxious boy's, and Johnny frisked away as if he were the sole giver of the feast.

Later that evening the visitors had all gone, the tea was all drunk, and the cake and the bread and honey were eaten. There was no sign left of such a great festival, except some freshly gnawed pickets in the front-yard fence where the horses had stood. Most of the guests had taken home with them a goodly piece of honeycomb, and there was not a great deal of honey left, but somehow nobody felt very sorry.

'I like to have company; don't you, father?' asked the boy.

He had a first-rate four-bladed knife in his pocket that the minister had given him. Johnny Hopper, though so wise and instructed a person, had never known before that the minister was such a nice man.

Mrs Hopper began to feel very tired. 'I thought one time, 'long at the first of it, they did look a little 'shamed, all of 'em meetin' here at once so, an' come for just what they were goin' to get,' she complained, fretfully.

'Land sakes, Ann Sarah, what's the use o' talkin' that foolish way?' said Grandma Prime, who was very social by nature and still abloom with happiness. 'You've al'ays got to have somethin' pleasant to draw folks round ye. I guess none o' them little bees won't think their labor was in vain in the Lord.* A nice afternoon like this ought to cost a little somethin', an' we've got some honey left.'

PHARMACOPOEIA

Jewett knew a good deal about flowers and plants, partly as a result of time spent with her father, who was a physician and amateur naturalist, and with her close friend Celia Thaxter. I am reprinting for reference here a pharmacopoeia compiled by Ted Eden for *The Country of the Pointed Firs* and originally published in *Colby Quarterly*, 28: 3 (September 1992), 140–3. I am grateful to *Colby Quarterly* for permission to reprint this list. I have added some of the less well-known plants mentioned in other stories, and a few other details.

Ash (tree) represents 'grandeur' in the nineteenth-century language of flowers.

Balm an herb used to soothe cuts and headaches.

Bay a leaf used for flavouring soups and sauces; used medicinally for indigestion, headache, and rheumatism; used to crown famous poets in ancient Greece (see 'laurel').

Bayberry a short, thick wild bush which grows along the coast in New England; the wax from its berries is used to make scented candles; represents 'instruction' in the nineteenth-century language of flowers.

Bergamot (also known as *bee balm*) used to soothe sore throats and as a mild sedative.

Bloodroot a plant with red root and sap, bearing a single white flower in early spring; related to the poppy; used to induce vomiting.

Borage used medicinally to reduce fevers, relieve liver and kidney troubles, as a laxative; also used in salads.

Camomile an herb used to induce sweating (for treatment of fever), for upset stomachs and indigestion, as an eye bath to refresh eyes, for cramps, and as a tonic (a kind of general pick-me-up).

Cardinal flowers a North American lobelia with a spike of brilliant red flowers. Represent 'distinction' in the nineteenth-century language of flowers.

Catnip used (paradoxically enough) to soothe people who have fits.

Cinnamon roses a species of rose (*R. cinnamonea*). Biographer Francis Matthiessen reports that as a child, Jewett would make a coddle of cinnamon rose petals with cinnamon and brown sugar. See *Sarah Orne Jewett*, chapter 1.

Clustering mallows (also known as *mallows* and as *marshmallows*) to relieve coughs and bronchitis, to soothe skin inflammations, to relieve sore throat; represent 'sweet disposition' in the nineteenth-century language of flowers.

Elecampane coarse herb; used to make sweetmeats (which are something like nuts) and for sore muscles.

Fairy-fingers local name for fox-glove, a tall plant with purple or white flowers that look like glove fingers.

Hyssop an herb used to make holy water (in the Catholic Church) and as a charm; used medicinally for colds, coughs, catarrh; as an expectorant (to clear out lungs and mucous membranes); for bruises and sprains.

Jerusalem cherry-tree either of two kinds of ornamental house plant with red or orange berries; nightshade family.

Laurel (also known as *bay laurel*) a slow-growing tree whose leaves were used in ancient Greece to crown victors in athletic contests and as a sign of honorary office; used in sauces and soups.

Lemon balm used as a tonic and to relieve headaches.

Lilac a common garden bush with heart-shaped leaves and strong-smelling purple blossoms; blooms early in the spring; signifies a first expression of love or 'forsakenness' in the nineteenth-century language of flowers; used by Walt Whitman as a symbol of the renewal of life through death in his elegy for President Lincoln, 'When Lilacs Last in the Dooryard Bloom'd'.

Linnaea (also known as *twin-flower*) a delicate wild plant difficult to grow outside of its native habitat.

Lobelia represents 'arrogance' in the nineteenth-century language of flowers.

Marigold represents 'jealousy' or 'uneasiness' or 'despair' in the nineteenth-century language of flowers.

Marsh rosemary a wild herb used to improve circulation and to relieve nervous headaches; also applied to insect stings and bites.

Mayflower any of several spring-blooming plants, such as the trailing arbutus, the hepatica, or various species of anemone.

Mint for flavour; mint tea relieves morning sickness during pregnancy.

Mullein an herb used for colds and coughs.

Pennyroyal an herb used for mosquito repellent, to improve digestion and relieve flatulence, for bronchial ailments, to purify water, and to promote expulsion of the placenta in childbirth.

Portulaca (also known as *purslane*) a brightly coloured, five-petal flower used in salads and as a tonic.

Ragged sailors name given to several kinds of flowers: bluebottle, bachelor's buttons, cornflower, prince's feather.

Sage an herb used to flavour meats; as a hot drink for coughs, colds, and constipation, as a tonic tea; represents 'wisdom' in the nineteenth-century language of flowers.

Simple an old word for an herb.

Snowberry related to honeysuckle; with white berries.

Southernwood used to make beer; also used as an antiseptic and stimulant; said to cure baldness.

Spice-apples early variety of ordinary apples.

Sweet-brier a European rose with big, strong thorns and a single pink blossom; rose hips (the fruit) are a source of vitamin C and are used for jellies and sauces; stands for 'poetry' in the nineteenth-century language of flowers.

Sweet fern sweet or aromatic North American shrub.

Sweet-mary sweet marjoram, an herb used for cooking; as a mouthwash and gargle for sore throat; for rheumatism; used as snuff to relieve headaches.

Tansy an herb with an aromatic odour and very bitter taste; used as a tonic; name comes from the Greek word for immortality; stands for 'resistance' in the nineteenth-century language of flowers.

Thoroughwort an herb used to induce sweating (for treatment of fever).

Thyme an herb used for cooking; as an antiseptic rinse for cuts; to treat asthma; for clearing up skin spots and pimples; symbolic connotations in the nineteenth-century language of flowers include strength, happiness, remembrance, time, and virginity.

Windflower an anemone, a flower with no petals but brightly coloured sepals; in the nineteenth-century language of flowers anemones represented sickness.

Witch-hazel a fall-blooming shrub with thin-petalled yellow flowers, the bark of which is used in a soothing but astringent lotion.

Wormwood an herb used medicinally and to make absinthe (an alcoholic liqueur); symbolizes gall and bitterness; stands for 'absence' in the nineteenth-century language of flowers.

Yarrow used as a diuretic and to treat fever, rheumatism, and flatulence; used to stimulate hair growth; made into a lotion to cleanse skin and to heal cuts and burns; chewed to relieve toothache.

EXPLANATORY NOTES

The Country of the Pointed Firs

3 *Alice Greenwood Howe*: a close friend and sometimes neighbour of Jewett, Mrs Howe expressed great admiration for *The Country of the Pointed Firs* when it first appeared in serial form in the *Atlantic Monthly*. According to Elizabeth Silverthorne, Jewett's dedication was a surprise to her friend. See *Sarah Orne Jewett: A Writer's Life*, chapter 10.

10 *cunner line*: a cunner is a small, salt-water food fish found along the coast of eastern North America.

17 *'A happy, rural seat of various views'*: John Milton, *Paradise Lost* (1667), iv. 247.

18 *Countess of Carberry*: Frances, the second wife of Richard Vaughan, second Earl of Carberry, who was well known for her piety in seventeenth-century England.

Darwin . . . schoolmaster!: Jewett appears to have misremembered or modified a passage from Charles Darwin's (1809–82) letter to Charles Whitley of 23 July 1834, which appears in *Life and Letters of Charles Darwin* (1887), edited by Darwin's son, Francis.

21 *Solomon's Temple*: see 1 Kings 6.

23 *Parry's Discoveries*: William Edward Parry (1790–1855) was an English explorer who made three voyages into the Arctic in search of the Northwest Passage. The Parry Islands were named after him.

24 *Man cannot live by fish alone*: Jesus admonishes the Tempter that 'Man shall not live by bread alone.' See Matthew 4: 4 and Deuteronomy 8: 3.

25 *Fox Channel*: Foxe Channel connects Hudson Bay with Foxe Basin, west of Baffin Island in the Northwest Territories of Canada.

27 *raised incessant armies*: this phrase and several lines in Littlepage's following speech come from or paraphrase Milton's *Paradise Lost*, vi. 138, 235–6, 242–4, a passage recounting a battle between the Heavenly Hosts and Satan's rebel angels.

Ge'graphical Society: the Royal Geographical Society, which was founded in London in 1830, in part to encourage exploration.

28 *recent outlines*: most of the Arctic regions had been mapped thoroughly by 1879, when the Northwest Passage was first navigated.

29 *shipping is a very great loss*: the Embargo Act of 1807, sometimes known as Jefferson's Embargo, devastated the smaller New England ports. Its purpose was to punish England and France for capturing neutral ships and impressing sailors for use in their fiercely contested war, but it was a costly, much-resented strategy.

30 *sleevin'*: hanging upon someone else's sleeve for assistance; probably intended ironically here.

32 *liable*: likely.

34 *them as fetch a bone'll carry one*: a variation on the popular proverb: 'The dog that fetches will carry.'

38 *ripped an' sewed over two o' them long breadths*: in the nineteenth century, carpet was manufactured in narrow widths (36-inch and 27-inch—90 cm. and 67.5 cm.—widths were typical) and often could be used on both sides. Turning a carpet might involve removing tacks from the edges, ripping seams where widths had been sewn together, beating and airing the pieces outdoors, replacing straw or matting used as padding, and then reassembling the pieces, with the more worn surfaces turned downwards or placed in the less travelled part of the room.

a large figure of Victory: Jewett could have seen the headless eight-foot-high Greek marble *Victory of Samothrace* (200–190 BC) at the Louvre in Paris.

39 *crystallized bouquets*: decorative bouquets of grasses and flowers can be made by coating them in crystals, usually sugar, though in this case salt would be the more likely choice.

46 *just as they be*: Jewett often repeated her father's advice concerning her own writing, to tell things 'just as they are'.

47 *right pattern of the plant . . . imitation*: an allusion to the philosophy of forms of Plato (428–348 BC); see especially *The Republic*, book vi.

48 *Antigone alone on the Theban plain*: see *Antigone* by Sophocles (496–406 BC), in which Antigone, daughter of the blinded and exiled King Oedipus of Thebes, defies the new King Creon's

orders and buries her rebellious brother after he dies in battle on the Theban plain.

49 *Tobago*: an island of the Lesser Antilles, off the coast of Venezuela.

50 *Sweet Home*: 'Be it ever so humble, there's no place like home.' This line from the chorus of 'Home, Sweet Home', the popular song from *Clari, The Maid of Milan* by John Howard Payne (1791–1852), conveys its main sentiment.

Cupid an' the Bee: possibly a musical setting of the fourth poem of 'Anacreontics' by Edmund Spenser (*c.*1552–99). When Cupid is stung by a bee, his mother suggests that he be less cruel in wounding lovers in the future, but he soon forgets the sting and resumes his normal behaviour.

52 *after the fashion of Queen Elizabeth*: Elizabeth I of England (1533–1603) reigned from 1558 until her death. Her frequent Royal Progresses were one means of sustaining her authority.

54 *child who stood at the gate in Hans Andersen's story*: Hans Christian Andersen (1805–75) was a popular Danish writer of novels, stories, and fairy-tales. The figure of an excluded child waiting at a gate appears in several of his stories, including 'What the Moon Saw' (Nineteenth Evening) and 'Lucky Peer'.

55 *an idyl of Theocritus*: Theocritus was a Hellenistic Greek poet in about 270 BC. He is credited with inventing pastoral poetry.

59 *bangeing-place*: a place to idle or loaf. The word 'bangeing' is also used to mean taking advantage of someone's hospitality.

69 *voice o' God out o' the whirlwind*: see Job 38: 1.

70 *the unpardonable sin*: two biblical passages refer to the unforgivable sin of blaspheming the Holy Ghost: Matthew 12: 31–2 and Mark 3: 29. Hebrews 6: 4–8 and 10: 26–31 suggest that Christian believers who lose their faith cannot repent. See also 'Ethan Brand' by Nathaniel Hawthorne (1804–64).

73 *saints in the desert*: according to tradition, St Anthony was the founder of the 'desert fathers', Christian hermits of the fourth-century church in Egypt.

75 *a place remote and islanded*: seems to echo Ishmael's affirmation of a central and insulated self that is a source of human happiness in Herman Melville's *Moby-Dick* (1851), chapter 58.

78 *tires want settin'*: wagon wheels had iron or steel tyres that loosened with use and had to be reset.

hearts and rounds: Francis Matthiessen reports that in her childhood, Jewett would visit her mother's elderly aristocratic friend Miss Cushing, who would serve her wine and small pound-cakes baked in hearts and rounds. See *Sarah Orne Jewett*, chapter 1.

84 *Entertainment for man and beast*: a version of a phrase common in advertisements for inns and hotels when travel was mainly by horse and on foot.

85 *its own livin' spring*: see Psalm 1: 3 and St Teresa of Avila (1515–82), *Interior Castle*, 'Fourth Mansions', chapter 2 and 'Seventh Mansions', chapter 2.

88 *bound girl*: an indentured servant.

90 *Pilgrim's Progress*: this popular allegory by John Bunyan (in two parts, published in 1678 and 1684) includes a number of processions, grand and small. One is the procession of angels that welcomes the transformed Christian and Hopeful into heaven at the end of Part I: 'Now I saw in my dream that these two men went in at the gate; and lo, as they entered, they were transfigured, and they had raiment put on that shone like gold.'

master afraid: extremely afraid.

91 *the god of harvests*: Demeter, the sister of Zeus and mother of Persephone, was the Greek goddess of the harvest. Associated with sacred groves, she was also the goddess of peaceful living in settled communities, of marriage, and of the family circle.

93 *Waterloo*: in this battle in Belgium (1815), Napoleon suffered his final military defeat.

Bunker Hill: in this American Revolutionary battle (1775)—actually fought at Breed's Hill, near Charlestown, Massachusetts,—even though American forces were driven from the hill, their fierce resistance prevented the British from breaking the American siege of Boston.

Decoration Day: celebrated on 30 May to honour the dead of the American Civil War by decorating their graves. It later became Memorial Day, on which all deceased American veterans are remembered.

94 *church military*: church militant; the Christian Church on earth seen as at war with the powers of evil, an aspect reflected in the popular hymn 'Onward, Christian Soldiers'.

96 *below the salt*: customarily, at table in medieval manor-houses, a salt-cellar, containing the comparatively expensive commodity, divided the family and friends, who sat 'above the salt', from the servants and dependants, who sat 'below the salt'.

100 *great national anniversaries*: for the International Centennial Exhibition of 1876, see note to p. 214. The World's Columbian Exposition of 1893, celebrating the voyage of Columbus to America, took place in Chicago.

109 *ary*: any.

   shortening the very thread of time: in Greek mythology, Clotho, Lachesis, and Atropos controlled human destiny, spinning the threads of individual lives and cutting them off at death.

The Queen's Twin

121 *Land of Eshcol*: see Numbers 13.

122 *the sea-serpent or the lost tribes of Israel*: see Isaiah 27: 1 for the sea serpent. Ten tribes of Israel rebelled against the House of David, leaving only Judah and Benjamin in the Kingdom of Judah. There has been considerable speculation about the identities and later locations of the lost ten tribes.

123 *the Evil One'll*: Satan; the devil.

124 *born the very same day*: Alexandrina Victoria was born on 24 May 1819 and was Queen of England and Ireland from 1837 until her death in 1901. She married Prince Albert of Saxe-Coburg-Gotha in 1840 and had nine children. The prince died in 1861. Queen Victoria's first Jubilee was held in 1887 to celebrate the fiftieth year of her reign; a second one was held in 1897. Her two books on the highlands are *Leaves from a Journal of our Life in the Highlands 1848–1861* (1868) and *More Leaves* (1883).

A Dunnet Shepherdess

141 *Medea's anointing Jason*: to gain the Golden Fleece in the Greek legend, Jason had to encounter fire-breathing bulls. Medea, later his queen, provided a magic ointment to protect him from the flames.

   minnies: minnows.

142 *like Macbeth*: see William Shakespeare's *Macbeth* (1606), e.g. Act III, Scene iv and Act V, Scene v.

149 *"Shepherd o' Salisbury Plain"*: a Christian religious piece published in 1795 by Hannah More.

   our Bibles . . . shepherd: that sheep are in need of a shepherd is a repeated theme in the Old and New Testaments of the Bible. See e.g. Psalm 23 and John 10: 11–14.

150 *like a figure of Millet's*: the French painter François Millet strongly influenced Jewett and her artistic friends. See e.g. Paula Blanchard, *Sarah Orne Jewett: Her World and her Work*, chapter 17. Millet produced several paintings of shepherds and shepherdesses that Jewett might have seen in collections in Paris, Boston, or Washington, DC or in reproductions. Among them are 'The Shepherdess Guarding her Flock' (1862–4) and 'Falling Leaves' (1866–7).

153 *Chaucer's time*: a reference to Geoffrey Chaucer (c.1342–1400).

154 *fled faster than Atalanta*: abandoned by her father, Atalanta was nursed by a bear and brought up by hunters to be a huntress. The man who wanted to marry her had to outrun her in a race. See *Metamorphoses* by Ovid (42 BC–AD 18), book x.

156 *Jeanne d'Arc*: Joan of Arc (c.1412–31).

The Foreigner

160 *rote*: the sound of surf on the shore.

162 *line storm*: an equinoctial storm. See John Greenleaf Whittier's *The Palatine* (1867).

163 *highbinders*: rowdies; a gang that commits outrages on persons or property.

164 *kick-shows*: kickshaws, trifles.

167 *Washin'ton pie*: a dessert made by alternating layers of cake with fruit jam (or jelly) filling.

168 *David's dancin' before the Lord*: see 2 Samuel 6: 14.

Orthodox: Paula Blanchard indicates that for Jewett's Maine readers, this term would mean 'Congregationalist'. See *Sarah Orne Jewett: Her World and her Work*, 320.

169 *stranger in a strange land*: see Exodus 2: 22. A similar phrase appears in Sophocles' play *Oedipus at Colonus*.

171 *officious*: Mrs Todd uses the word in its archaic sense, meaning 'helpful', 'obliging'.

172 *Sir Philip Sidney's phrase*: Sir Philip Sidney (1554–86) was the author of the *Arcadia* (1581), now known as the *Old Arcadia*, and the *New Arcadia* (c.1583–4), from which this quotation comes in book i; 'made' is a variant of 'mad' in some MSS.

173 *fête day*: the feast-day of the saint after whom one is named, celebrated in some countries, such as France, like a birthday.

175 *New Jerusalem*: see Revelation 21: 2.

181 *statue of the Empress Josephine . . . in Martinique*: Empress Josephine was the wife of Napoleon Bonaparte (1769–1821). Martinique is a French island colony of the Lesser Antilles, between Puerto Rico and Venezuela.

sibyl of the Sistine Chapel: Michelangelo (1474–1564) placed representations of sibyls around the nine frescoes on the ceiling of the Sistine Chapel of the Vatican that depict Old Testament stories from the Creation to the story of Noah.

184 *we've got to join both worlds together an' live in one but for the other*: see Romans 12: 1–2 and Ephesians 4: 23. According to Paula Blanchard, Jewett believed that communication can occur with the dead who live in a spirit world that watches over our own. In *Sarah Orne Jewett: Her World and her Work*, chapters 15 and 20, Blanchard reports that Jewett's close friend Celia Thaxter, shortly before her death in 1894, saw a vision of her mother at the foot of her bed. See also Blanchard's discussion of the death of Jewett's father in chapter 10 and F. O. Matthiessen's discussion of the death of her mother in *Sarah Orne Jewett*, chapter 3.

William's Wedding

185 *Carlyle says somewhere . . . conspiracy*: Thomas Carlyle (1795–1881) expressed this general idea in several of his works, including 'Characteristics' (1831) and *Sartor Resartus* (1833). The passage to which Jewett refers is most likely *Past and Present* (1843), iii, chapter 4.

187 *line-storm*: see note to p. 162.

189 *Santa Teresa says . . . loving*: see St Teresa of Avila, *Interior Castle*, 'Fourth Mansions', chapter 1.

'The happiness of life is in its recognitions . . . to us—': Jewett implies that this quotation also is from St Teresa, but this has not been established.

191 *Medea could not . . . ultimate purposes*: see note to p. 141.

down from the mount . . . so bright: an allusion to the transfiguration of Christ. See Matthew 17: 2 and Mark 9: 2.

192 *all their ideas come down from Sodom*: see Genesis 19: 28 and also Genesis 14: 1–4.

195 *the kiss of peace*: this was one of the rites of the Eucharistic service in the primitive Christian Church. See Romans 16: 16.

Fair Day

201 *lottin' on*: depending on, expecting.

204 *lump of beeswax*: beeswax was used to coat thread to improve its strength and to ease the process of drawing it through multiple layers of fabric, as in quilting.

The Flight of Betsey Lane

214 *the Centennial*: during May–November 1876 an International Centennial Exhibition took place in Philadelphia to commemorate the signing of the United States Declaration of Independence in 1776.

215 *Callao*: the main seaport of Peru.

218 *feel cropin' about*: this apparently local usage does not appear in standard dictionaries. The context indicates 'a longing for', 'a lack of'. The word 'croping' may be related to 'groping', in the sense of feeling around for something one cannot see.

219 *below the salt*: see note to p. 96.

223 *begretched*: begrudged.

224 *pewee*: name given to several types of small fly-catching birds.

226 *day o' judgment*: see Revelation 20: 11–14.

229 *beakin'*: probably refers to head movements such as those a bird makes to see objects on either side of its beak; see p. 230: 'moving her small head from side to side'. The *Oxford English Dictionary* notes a similar usage: 'to push the beak (or snout) into'.

232 *Sindbad*: this man from Baghdad has adventures at sea in *The Arabian Nights' Entertainments* (*The Thousand and One Nights*).

The Only Rose

241 *meechin'*: meek, self-effacing. According to Paula Blanchard, this term had a range of negative connotations. See *Sarah Orne Jewett: Her World and her Work*, chapter 18.

245 *I'm out o' my time*: John means that he has completed an apprenticeship.

The Guests of Mrs Timms

248 *county conference*: a meeting of representatives from local churches. See p. 253.

250 *turned again*: a dress can be remade by taking it apart—and usually cleaning it—turning it inside out, and putting it together again.

251 *drop-cakes*: small cakes made by letting batter drop from a spoon into hot fat or onto a greased pan to be baked in an oven.

252 *'When it rains porridge hold up your dish'*: Jewett's own version of a fairly well-known saying regarding the fool or unfortunate who, when it rains porridge, has no bowl or spoon. See for example T. C. Haliburton's *Wise Saws* (1843), chapter 27.

253 *Orthodox . . . Freewill Baptists*: the Free Will Baptists were a Protestant sect founded by Benjamin Randall in 1780 in New Durham, Massachusetts. A central tenet of their doctrine was that God's grace was free to all who would embrace it. This democratic theology contrasted with the more hierarchical Calvinist theology of the Orthodox or Congregationalists, the American Church founded by the Puritans who colonized Massachusetts and southern Maine.

256 *more generous with her little than many was with their much*: possibly an allusion to the poor but generous widow in Mark 12: 42–4 and Luke 21: 1.

261 *'Better is a dinner of herbs where love is'*: see Proverbs 15: 17.

A Neighbor's Landmark

264 *sprangly*: probably from 'sprang', a weaving technique in which the threads are twisted over each other to form an open mesh. 'Sprangly' would refer, then, to the pine with the drooping boughs described on p. 277 below.

spile: spoil.

268 *what it says in the Bible about movin' a landmark*: there are several passages in the Bible that forbid the moving of landmarks. See Deuteronomy 19: 14; 27: 17; Proverbs 22: 28; 23: 10.

273 *ellums*: elms.

275 *rote*: the sound of surf on the shore.

The Life of Nancy

281 *the Common*: Boston Common, a 48-acre tract originally reserved in 1634 as pasture and training-field.

283 *Rip Van Winkle . . . this week*: Joseph Jefferson and Dion Bouci-
cault wrote a popular dramatization of 'Rip Van Winkle', Washington Irving's well-known story from *The Sketch Book*
(1819–20). The play was first performed in 1865, and Jefferson
was enormously successful in the role of Rip.

horse-car: horse-drawn trolley, streetcar, or tram.

284 *Missionary Building . . . Somerset Street*: there was a Missionary
House at 33 Pemberton Square; the building extended back to
Somerset Street. It housed the offices of the American Board of
Commissioners for Foreign Missions, whose purpose was 'to
send out consecrated men to all parts of the world and establish
schools, churches and powerful native ministries'. The *Missionary Magazine* had its offices at 33 Somerset Street.

Faneuil Hall: this two-storey building with a bronze dome is
called 'the Cradle of Liberty' because mass meetings were held
there in the years before the American Revolution.

Old South Church: the Old South Meeting House, home of a
Congregational church, was the site of pre-revolutionary town
meetings, including the one that began the Boston Tea Party. In
1875 the congregation erected a new building, which became
known as 'New' Old South Church.

285 *Harvard assembly*: that Tom has attended an assembly suggests
that he is a member of one of the exclusive Harvard clubs and,
therefore, ranks high in Boston society. Ronald Story reports in
The Forging of an Aristocracy (Middletown, Conn.: Wesleyan University Press, 1980) that 'Boston Assemblies' were held beginning in the 1830s at Papanti's ballroom (see note to p. 289
below), providing an opportunity for Harvard students to social-
ize (p. 94).

287 *Great Organ*: almost certainly the organ in King's Chapel, which
was installed despite local opposition to having such an instrument in a church. From the late eighteenth century the organ
was used for concerts.

the monument: Bunker Hill Monument in Charlestown, a 221-
foot granite obelisk, with a public spiral staircase that one can
ascend for the view. See note to p. 93.

289 *Papanti's*: though opposed by local clergy, Lorenzo Papanti
founded a dancing academy in Boston which became an institution. In *Boston and the Boston Legend* (Boston: Appleton-Century,
1935), Lucius Beebe writes, 'All good Boston children went to
Papanti's, where his lean figure, glossy wig and elegant patent

leather dancing pumps, and above all his pointed fiddle-bow, used both as an instrument of correction and harmony, struck terror to all juvenile hearts' (p. 39).

295 *Lancers quadrille*: a type of square dance of French origin.

296 *David, he danced afore the Lord*: see 2 Samuel 6: 14.

298 *seed fell upon good ground*: see Matthew 13: 8.

Martha's Lady

302 *Dr Johnson*: Samuel Johnson (1709–84).

304 *poems of Emerson*: Ralph Waldo Emerson (1803–82) published his first collection of poems in 1847.

309 *White Hills*: the White Mountains of northern New Hampshire; see note on *the new Profile House* below.

310 *the new Profile House*: a popular White Mountain resort hotel in Franconia Notch, New Hampshire.

314 *Boston Common*: see note to p. 281.

Bold Words at the Bridge

320 *clat'*: probably clatter.

   Faix: faith.

   dwining: dwindling.

321 *ates*: probably 'eats'; see below, p. 326.

   whillalu: hullabaloo.

322 *Wisha!*: an expletive, like 'Oh well!' or 'Pshaw!'

324 *County Kerry*: in Ireland.

   Bad cess: bad success, or bad luck.

326 *lately come out*: recently arrived from Ireland.

   Cork . . . Bantry . . . Kinmare: towns in Ireland.

327 *stramming*: huge.

The Honey Tree

330 *Sho! sho!*: sure, sure.

   trollick: this apparently local usage or coinage does not appear in standard dictionaries. The context makes clear that it means extraneous matter or trash. The word may be related to the

Scottish use of 'troll' and 'trollop' to refer to unshapely and slovenly things, trailing on the ground or hanging wet.

331 *harness-cask*: a cask or tub in which salt meat is stored for daily use (nautical terminology).

332 *bate*: probably the Maine pronunciation of 'bet'.

336 *Marathon*: at the Battle of Marathon in 490 BC, Athenians and Plataeans attacked and defeated the more numerous Persian invaders, thus avoiding Persian domination.

338 *gybe*: variant of 'jibe', meaning to be in accord or agree.

340 *in vain in the Lord*: see 1 Corinthians 15: 58.

THE WORLD'S CLASSICS

A Select List

HANS ANDERSEN: Fairy Tales
Translated by L. W. Kingsland
Introduction by Naomi Lewis
Illustrated by Vilhelm Pedersen and Lorenz Frølich

ARTHUR J. ARBERRY (Transl.): The Koran

LUDOVICO ARIOSTO: Orlando Furioso
Translated by Guido Waldman

ARISTOTLE: The Nicomachean Ethics
Translated by David Ross

JANE AUSTEN: Emma
Edited by James Kinsley and David Lodge

Northanger Abbey, Lady Susan, The Watsons,
and Sanditon
Edited by John Davie

Persuasion
Edited by John Davie

WILLIAM BECKFORD: Vathek
Edited by Roger Lonsdale

KEITH BOSLEY (Transl.): The Kalevala

CHARLOTTE BRONTË: Jane Eyre
Edited by Margaret Smith

JOHN BUNYAN: The Pilgrim's Progress
Edited by N. H. Keeble

FRANCES HODGSON BURNETT: The Secret Garden
Edited by Dennis Butts

FANNY BURNEY: Cecilia
or Memoirs of an Heiress
Edited by Peter Sabor and Margaret Anne Doody

THOMAS CARLYLE: The French Revolution
Edited by K. J. Fielding and David Sorensen

TOBIAS SMOLLETT: The Expedition of Humphry Clinker
Edited by Lewis M. Knapp
Revised by Paul-Gabriel Boucé

ROBERT LOUIS STEVENSON:
Treasure Island
Edited by Emma Letley

ANTHONY TROLLOPE: The American Senator
Edited by John Halperin

GIORGIO VASARI: The Lives of the Artists
Translated and Edited by Julia Conaway Bondanella and Peter Bondanella

VIRGINIA WOOLF: Orlando
Edited by Rachel Bowlby

ÉMILE ZOLA: Nana
Translated and Edited by Douglas Parmée